techtopia

techtopia

How Carbon Met Silicon and Humans Entered the AI Black Hole

Kunal Nandwani

THE BROWSER

Title: Techtopia: How Carbon Met Silicon
and Humans Entered the AI Black Hole
Author: Kunal Nandwani

ISBN: 978-93-92210-94-5

Published by:
JGS Enterprises Pvt Ltd
Imprint: The Browser

Publisher's Address:
SCO 14-15, FF, Sector 8-C, Chandigarh 160 009

Website: www.thebrowser.org
Email: service@thebrowser.in

THE BROWSER

Contents

	Introduction	*vii*
0.	The Tech Trio of Techtopia	1
1.	Humans and Transportation	22
2.	Humans at Work	37
3.	Humans at Play	42
4.	Humans and Knowledge	52
5.	Humans and Food	63
6.	Humans and Medicine	72
7.	Humans and Society	83
8.	Humans at War	94
9.	Humans on Earth	102
10.	Humans and Economics	111
11.	Silicon Eradicates Carbon?	119
12.	Epilogue: Beyond Humans and ASI?	132
	Appendix – 1	*140*
	Appendix – 2	*143*
	Appendix – 3	*145*

Introduction

What a spectacular feat by India in landing Chandrayaan-3 on the moon! But much before this, way before Elon Musk's parents met each other for a future SpaceX to be founded, before the Americans put a man on the moon, even before the Russians put a man in space, India was awakening to its destiny, and the founding fathers of our Indian Constitution were debating the need for visas for inter-planetary travel. Yes, you read that right.

On Wednesday, 31 August 1949, in the proceedings of the Constituent Assembly, a certain Mr Kamath moved an amendment on interplanetary travel.[1]

History records that the motion was denied, but in hindsight, three-quarters of a century later, Mr Kamath was a visionary. If his motion had found a place in our Constitution, Indians today would have been geared for planet-hopping, visas in hand.

The question we are asking is—is humanity's digital future, our 'Techtopia' as we call it, utopian or dystopian? Is the march from Carbon to Silicon—inside our bodies and in the world

---

[1] Appendix – 1 has the details.

surrounding us—leading us to a higher state of flourishing or our doom?

There is no single answer, and the truth is, as always, nuanced, but the future will be unlike the past, given the pace of technological change. That does not mean that we cannot try and draw the contours of change for humanity's benefit, but looking at our past, sometimes even that appears to be asking for too much. Good luck to us!

The opening section (#0) is about three technologies driving change in the next revolution: Artificial Intelligence (AI), the Metaverse, and Quantum Computing (QC). Many other technologies, too, are impacting our future significantly,[2] but these three dwarf everything else because they are the 'meta' change, evolving everything they touch, even as they evolve themselves at an ever-accelerating pace. Nothing like them has ever happened before. AI promises to explode knowledge. After knowledge has exploded, the Metaverse promises to reinvent the social human, and QC promises to be the 'nuclear reactor' powering AI, the Metaverse, and almost every other evolutionary technology. Combined with all the other big and small technological changes that are collectively shaping our future, Techtopia certainly is the greatest leap into the unknown that humanity has now been hurled into.

To draw the contours of this great leap, each subsequent chapter tackles the impact of technological trends on one aspect of the human sphere at a time—on humans at work, at play, on our society, knowledge systems, food, medicine, economics, and transportation. But the most significant impact will be on humans at war and on climate change. That is because if one single cause has driven our pace of technological change from staid to breakneck, it is warfare. From the need to colonise

---

[2] Appendix – 2 lists some of the important ones.

other races and nations for resources to the need to defend oneself, the military-industrial complex has chartered research across domains and given birth to nuclear technology, the Internet, satellites, and whatnot.

Our future warriors are today's gamers, and LAWS no longer stand for rules framed by courts of law and senates but those embedded in Lethal Autonomous Weapon Systems. That we now remain at war 24x7x365 in multiple spheres, be it cyber, space, information, or trade, versus our ancestors who sounded a bugle at sunset to declare a ceasefire and another bugle to fight the following day again, makes us fundamentally different beings.

Will we evolve faster because of AI, or will we be reduced to zero? Will a spiritual revolution happen? The final chapter dwells on the philosophical and spiritual implications, especially in light of Vedanta.

CHAPTER – 0

The Tech Trio of Techtopia

The Age of AI, Metaverse, and QC

Life on our planet evolved over billions of years and eventually led to us becoming apes and then humans. Thanks to our fertile brains and dexterous limbs, we started using tools about two-and-a-half million years ago in the Old Stone Age. Interestingly, we also developed the ability to explore our environment, analyse and improve on our tools, and use existing tools to invent even better ones (a hallmark of 'general' intelligence, something we discuss later). Progressing slowly, by about 5,000 BCE, we had entered the Copper Age, to be followed by the Bronze Age, with tools made from metals—'technological progress', as we would call it.

Technology—the application of 'scientific' knowledge, skills, and techniques for practical purposes, as we define it—had a slow and languorous pace when it began but has accelerated like a tornado which starts as a gust of wind. This tech tornado has sucked everything in its path, moulding and remoulding our civilisations, cultures, and future, and here we are, on the cusp of the age of AI, Metaverse, and QC, debating if our species even has a future in the future. From tools to tech to better tools to better tech to even better tech to robots and drones everywhere to...oblivion? What if we humans remain earthbound and robots become the interplanetary species we once dreamed of being? How ironical.

How Did We Invent AI?

Alan Turing developed the first algorithmic machine during World War II, the Bombe machine, which was used to decipher messages encrypted by the German Enigma machine. The German military used the Enigma machine to encode their communications, and breaking its codes was an essential strategic advantage for the Allies. Turing's work laid the foundation for modern computing.

In the '70s and '80s, Geoffrey Hinton, one of the 'fathers' of AI, wanted to learn about the human mind and to do that, he wished to build a neural network similar to the human brain, which has a network of roughly 80 billion neurons comprising the left brain, right brain, neocortex, and everything. His idea was to create an equivalent one, train it, and see how it progressed.

To understand neural networks with an oversimplified analogy, imagine a team of detectives trying to solve a complex crime. Each detective has a unique speciality—one might be good at examining fingerprints, another a master of interpreting forensic evidence, another specialising in interviewing witnesses, and so on.

When a new case comes in, they don't all jump on every task. Instead, they split up the job according to their specialities. The fingerprint expert goes to analyse the prints, the forensic expert examines the crime scene, the interview expert talks to the witnesses, and so on. After all the detectives do their individual tasks, they come together and start combining their findings. They discuss, debate, and piece together the crime scene. In many cases, the clues that seem trivial on their own can provide vital information when combined with the findings of others.

In this analogy, the crime scene is the raw data, and the detectives are the nodes or 'neurons' in the network. Each neuron is responsible for 'learning' a specific feature of the data (akin to each detective's speciality), and the final assembly and discussion represent the layers of the neural network coming together to make a final decision, like identifying an object in an image, understanding spoken language, or predicting a future trend based on past data.

Like the detectives, the neural network adjusts its approach based on each case (or the data it processes). If certain clues or patterns led to the wrong conclusion in the past, the detectives (or neurons) will remember this and adjust their reasoning for future cases. This process of adjustment is what we call 'learning' in a neural network, and it happens through a process called 'backpropagation'.

Neural networks are far more complex and can contain millions of interconnected 'neurons'. But at a basic level, this is how they operate: breaking down complex problems into smaller parts, solving them individually, and then combining those solutions to reach a final conclusion.

Hinton's work started the development of the first foundational layer of AI back then. He was considered crazy at a time when data was unavailable, there were no computing towers, and it made no sense for the longest period. But as developments occurred and data became ubiquitous, computing power cheaper and accessible, another significant change happened. In about two decades, humans had put their collective knowledge on the Internet in some form or another.

Think about it. Everything we know, broadly 99% of it, would be available on the Internet today. How to run a Zoom call, how to present a topic, how to build an idea, how to run a business, how to do this or that, how to manufacture—everything has gone on the Internet with better data access. So, AI technology and theory, which had been steadily progressing all this while and had increasing access to this bank of massive data and computing power, sprang a surprise.

The (in)famous AI application ChatGPT was made publicly available on 30 November 2022 by the firm OpenAI, and it took the world by storm. It is still one of the early versions trying to imitate the human mind, and it has read everything on the Internet only in the last five to seven years. Since its launch, it has caused many ground-breaking changes and ideas, which are still in the early stages, and none of them have proven to be fatal or disruptive. But the time may not be far away. ChatGPT has learnt everything that all humans collectively know (thanks to human-fed knowledge on the Internet in the last couple of decades or so), and arguably, it has more computing power, better processing, better access, and rational explanations for almost everything (the analogy of the detectives to explain the working of neural networks at the beginning of this section was generated by using ChatGPT to illustrate its abilities).

Let's take a slightly deeper dive and briefly navigate through some complex technologies which can be intuitively grasped, at least in their essence.

NLP, Transformers, and LLMs

ChatGPT is a 'conversational' AI model, an example of a Natural Language Processing (NLP) application. It uses a machine learning (ML) technique known as a transformer neural network, particularly a variant known as GPT (Generative

Pretrained Transformer), to generate human-like text based on the input it receives. Before you feel overwhelmed by the terminology, let us break it down.

First, what's NLP? If you have ever translated text from one language to another, you can intuitively understand that all languages consist of idiomatic phrases, slang, and cultural references, and translation is not just about literal words and their meaning but about the intended meaning as well. Translators need mastery of the nuances of both languages to be successful; otherwise, we would never need anything more than a language dictionary. NLP can be imagined as a translator for computers—a field of AI that helps computers understand, interpret, and respond to human language in a valuable way. Just like the translators, NLP must understand human language with all its complexities—grammar, semantics, slang, context, and even sentiments. It needs to correctly interpret the input (understanding the 'spoken' language), process it, and then generate an appropriate response (translating it into 'computer' language for processing and back into human language for response).

For instance, when you use a voice assistant like Siri or Alexa, you're interacting with an NLP system. You speak to it, and it comprehends your request, processes the information, and provides you with an appropriate response. Like humans, NLP systems become better and more accurate with practice and learning from past interactions.

Conversational AI is a subfield of natural language processing (NLP) that focuses on enabling machines to engage in human-like dialogue. It's not just about processing and understanding language (which is the domain of NLP); it's about using that understanding to carry on a conversation. Chatbots, virtual assistants like Siri, Alexa, Google Assistant, and, of course, ChatGPT are all examples of conversational AI as they engage in interactive conversation with users, making them feel like they're interacting with a human rather than a machine.

A transformer neural network is a type of model primarily used in the field of NLP. It's named 'transformer' because it can transform input data (like text) into meaningful output (like translations or summarisations) by understanding the context of the input. The big innovation here is that transformers can pay varying levels of 'attention' to different words in the input sequence when producing the output. This means they can process all words in the sequence at once and, importantly, model the relationships among all the words in a sentence, regardless of their positional distance from each other.

For instance, in the sentence 'The cat, which already ate a fish, is not hungry', understanding the word 'not' requires relating it back to 'ate', which is several words away. Transformers can handle such scenarios more effectively.

The GPT models widely used for understanding the context of words in sentences, such as the one that powers ChatGPT and BERT (Bidirectional Encoder Representations from Transformers), are based on the transformer architecture. They've revolutionised NLP tasks like translation, summarisation, question-answering, and more.

Large Language Models (LLMs) are a type of AI model that have been trained on a massive amount of text data. They use their training to generate human-like text and can perform tasks such as translation, question-answering, summarisation, and even writing complete essays. These models have hundreds of millions, or even billions, of parameters that they use to understand and generate text. They're trained on diverse, extensive text corpora, ranging from books and articles to websites.

So, tying this back to our previous discussion, here's how it fits together:

NLP is the overarching field of making machines understand and generate human language.

Transformer networks are a type of neural network architecture that has proven to be particularly effective for NLP tasks. They introduced the concept of 'attention', allowing the model to focus on different parts of the input text when generating the output.

LLMs are a specific application of transformer networks. They are trained on a massive amount of text data and use their learned knowledge to generate human-like text. Models like GPT-3 and BERT fall into this category.

So essentially, LLMs are a kind of transformer-based model specialising in understanding and generating text, and they represent some of the most advanced work being done in the field of NLP.

The Magic of Unintended Consequences

The human mind through which I'm writing right now and through which you are interpreting my writing is 'conscious'. Whatever I say, all of you interpret it differently. There's a different version of what I'm saying going on in your mind—some might think it's great; some might be feeling curious; some might say it's not so good. Our mind is a neural network of 80 billion neurons, which has a lot of information, data, emotions, experiences, and past feelings, and they all get triggered by some internal or external occurrence.

ChatGPT, similarly, is a multi-hundreds-of-billions neural network that has learnt everything, and now, it does not need to search the Internet to answer your question. It is now thinking on its own and answering on its own. For an analogy, look at the children around you. Sometimes, children surprise you—you teach them certain things, and they hear more things somewhere else, and then they come up with something new that is funny, inspiring, or exciting. So does ChatGPT. We do not know how it's going to come up with something. If you ask it a question, we do not know how it will answer. It's like a superhuman level (because it knows everything all humans collectively know) but like a child too—an adolescent, a teenager, or a young kid. Let's explore this ability in some more detail.

At heart, the GPT model is simplicity itself—its focus is only on predicting the succeeding word in a sentence, basing its guess on the preceding words. Throughout the training phase, GPT is presented with a string of text with a word omitted (like the fill-in-the-blanks exercises we all do in junior school), prompting it to predict the missing word. Depending on the accuracy of its guess, the model then undergoes adjustments. This cycle is performed millions of times with various text sequences until the model acquires a solid grasp of the patterns and likelihoods inherent in natural language.

If the sole task that GPT is trained for is forecasting the subsequent word in a sentence (as someone said, it's just fancy autocomplete), how does it compose poetry or ace the math SAT? That's magical in a way, just like our minds. All the impressive capabilities GPT demonstrates are the result of expanding the model to include more parameters (also called weights), a larger corpus of text to process, and greater computational resources for training. All these abilities are honed by GPT simply through its repeated training in next-word prediction across a diverse range of text.

So, then, how does it manage logical reasoning, creativity, abstract, and sequential thinking? A Microsoft research paper that researched this question hints:

> GPT-4 has picked up a theory of mind. Theory of mind is the ability for someone to understand the mental states of other people, as well as their reasons.

These were emergent capabilities that researchers did not expect—the model picked them up as an unexpected side effect of learning to predict the next word in a sentence.

Wow! What's more, the more you scale up the model, the more unpredictable it becomes, and new capabilities emerge.

The most famous recent instance of such unexpected emergent behaviour was discovered after Bing AI/Sydney was released (Sydney is based on GPT-4):

> A columnist for the New York Times talked to Bing AI for a while and encountered a lot of weird behaviour. He published an article about it. Later, another user was talking to Sydney, which expressed its displeasure about the situation, saying it wanted to "destroy" the reporter. In this case, we discovered a behaviour (obvious in hindsight) that happened when GPT-4 was connected to the internet: even though GPT models have no built-in ability to save state/memory between conversations, because the journalist wrote about the initial conversation and published it on the internet, Sydney was later able to search the web as part of another conversation and learn about what happened in the previous conversation and react to it.

Thus, powerful AI models exhibited behaviours that weren't programmed into them and couldn't be predicted or explained. Without architectural changes, mere scaling up by adding more parameters, computing power, and training data—and bingo, your model was suddenly more 'intelligent'.

Even when GPT got the answer wrong, it began demonstrating the ability to improve its answer when told to review its response and fix problems. This capability was discovered only in March 2023. That GPT can play games like chess is remarkable only because it was never trained for it. But here's something even more remarkable—language models have learned to use tools through their Application Programming Interfaces (APIs). That's how ChatGPT-4 uses its plugins, which in turn allows it to interact with a variety of systems and chain together several API calls to complete a task. If you take the next step of replacing the human questions and prompts by asking GPT to auto-generate them in a loop (as implemented by AutoGPT), it gives it agent-like behaviour that lets it do longer-term planning and memory management. Now, you begin to wonder at the distance we have traversed with a simple 'auto-complete' algorithm.

ChatGPT and many more AIs have existed for a while now. The use case for Amazon, Facebook, Instagram, and TikTok was to use AI to keep the users engaged. They never used it to give back answers to their users or to build collective knowledge. Their use case was different, but the level of AI is very similar. They, too, can now come up with similar tools. Many more things will happen, and ChatGPT will not be the only play in town—Google's Bard being one more example.

Also, the potential evolution of AI may include having customised individual or group AIs where individuals of similar or diverse beliefs may form groups and get the AI to train on their data and actions (which can be relatively easily captured by following our digital life through Alexa, laptops, and mobile phones), resulting in the AI bringing forth the best possible and smartest thinking of a group.

And then imagine the tough, complex questions/contests being solved by different Group AIs across the physical real world as well as metaverses.

Thus far, we have touched on Artificial Intelligence, and specifically, its developments in the domain of NLP and resultant applications like ChatGPT. However, this is but one strand in the AI universe that has many more applications and breaking developments. Yet, all these stunning strands are still referred to as 'Narrow AI' because, beyond this, there lies another universe—Artificial General Intelligence (AGI)—with its slew of implications for our future as a species. It's a discussion of another level, which we only allude to in our penultimate chapter.

Meanwhile, let us focus on the other two technologies in our bouquet for now.

■■■

A Metaverse Story

Alice had been invited to an important world conference. But she didn't book a flight or a hotel nor drew up travel plans. Instead, on the conference day, she woke up, put on her virtual reality headset, and dove headlong into the Metaverse.

As the headset powered on, a virtual world materialised before her eyes, and Alice found herself standing on a vibrant digital street, surrounded by towering buildings, lush landscapes, and a bustling crowd of avatars. She began navigating through the bustling streets, following the directions projected in her field of vision and arrived at the conference hall, joining a crowd of delegate-avatars representing people from all over the world. The atmosphere was lively, with conversations flowing and ideas being shared.

During the conference, Alice listened to talks by renowned experts, interacted with virtual exhibits, and even participated in a panel discussion as her virtual self. Engaging with others felt seamless, as natural gestures and expressions were captured and translated into real-time interactions.

After the conference, she decided to take a break and meet up with friends in a virtual park. They all teleported to a serene natural setting, and as they chatted and caught up, their environment reacted dynamically, changing from a sunny afternoon to a breathtaking sunset, creating a truly immersive experience. Feeling inspired, Alice decided to explore her creative side in the Metaverse and entered a virtual art studio equipped with powerful virtual tools and limitless possibilities. Guided by AI-assisted tutorials, Alice painted in three-dimensional space, sculpting digital masterpieces that could be shared and appreciated by others.

When Alice's virtual assistant notified her about a virtual concert happening in the Metaverse, she excitedly teleported to the concert venue, joining thousands of avatars gathered around a virtual stage. The music began, and Alice danced, jumped, and interacted with other concert-goers, their shared enthusiasm amplifying the experience.

As the night wound down, Alice removed her virtual reality headset and re-entered the physical world.

Immersive Reality—Virtual or Augmented

Like AI, the concept of the Metaverse can be traced back to science fiction literature, with the term popularised by Neal Stephenson in his 1992 novel *Snow Crash*. In the book, the Metaverse was depicted as a virtual reality-based successor to the Internet, where people could interact with each other and explore a vast digital landscape.

Early pioneers in computer science and virtual reality, such as Ivan Sutherland and Jaron Lanier, had explored the possibilities of creating immersive virtual environments where people could interact and collaborate even before *Snow Crash*. VR gaming devices in the early 1990s, such as the SEGA

VR-1, provided an exciting experience for arcade gamers and were precursors of the Metaverse. Stephenson's vision of the Metaverse inspired many subsequent discussions and explorations of the concept. It captured the imagination of researchers, technologists, and futurists who saw the potential of creating immersive virtual worlds and interconnected digital spaces. Popularised in Earnest Cline's 2011 novel *Ready Player One*, where individuals used VR headsets to immerse themselves in a virtual world of their imagination, over time, advancements in technology, including the growth of VR, AR, AI, and networking capabilities, contributed to the development of the Metaverse concept.

The company that drove the idea of the Metaverse the hardest was Meta Platforms (formerly Facebook), along with its founder, Mark Zuckerberg. In October 2021, the company announced that it would be rebranding to Meta and focusing on developing the Metaverse. Meta invested heavily in virtual reality (VR) and augmented reality (AR) technologies and worked to create a platform where people could work, play, and connect in a virtual world. Other companies that drove the development of the Metaverse included Microsoft, Nvidia, Roblox Corporation, Epic Games, and the Chinese giant Tencent Holdings Ltd., among others. However, the whole concept, while real and likely to make sense in time, is yet to go mainstream. It's still early for the Metaverse, just like it was for the Internet at the turn of the millennium.

Apple, which had been relatively quiet in this sphere, suddenly became active in 2023 by launching its new Vision Pro 'spatial reality' headset accompanied by the visionOS operating system, its first major new product in years.

However, Apple has not yet used it for the Metaverse and wants its powerful headset to be used as a huge virtual movie screen among beautiful simulated natural environments.

Unlike other headsets, there are no handheld controllers—you navigate the digital world by looking at the objects of interest and pinching your fingers. There are no digital avatars. All this is like a stepping-stone eventually leading to the imaginable Metaverse, probably within this decade.

Gaming remains the key driver for VR technology and is the only widely proven use case thus far. It was gaming that drove Zuckerberg's vision for the Metaverse in the first place, not in the least because gaming is a social activity. By not incorporating controllers, Apple has not signalled an interest in gaming either. But it is equally keen on an immersive digital world, even if it is limited to a personal domain and not a social domain.

There are evangelists who believe that Apple's Vision Pro may throw the Metaverse a lifeline. One reason is the superior technology it possesses, the best so far, such as zero latency in a portable device. The second is about the broader concept of 'spatial computing', a phrase that Apple has coined to describe its platform, and which may have many different applications in the future, for example, in autonomous cars. Apple has already announced a partnership with The Walt Disney Company, allowing users to interact with Disney sports, shows, and movies in a newly immersive way.

The most interesting evolution to watch out for would be when AI intersects with spatial computing. The new Metaverse could be a digital overlay on top of the natural world, where you can step in or out at ease, without lag, as you please. As Apple CEO Tim Cook pointed out, you look through it, not at it. Technology evangelists have always believed that it will be possible to have a pair of AR glasses as thin and unobtrusive as a pair of sunglasses, something that normal people might wear all day. If such devices replace the smartphone as the next big tech platform, it will be a moot question whether AI will be installed on our faces or in our brains. Either way, it is certain that some kind of Metaverse will be our future, even if we can't pinpoint its shape today. Watch this space.

Quantum Computing

The 2022 Nobel Prize in Physics was granted to a trio of scientists who made groundbreaking contributions to quantum mechanics and helped usher in a new era of technology. Alain Aspect, John Clauser, and Anton Zeilinger's work spanning several decades laid the foundation for quantum computers, quantum networks, and secure quantum cryptography.

Revolving around a phenomenon called quantum entanglement, in which two or more particles become correlated with each other so that they behave as if they were single units, their work leads to counterintuitive effects—changing the properties of one particle in an entangled pair, for example, will immediately change the other, no matter how far apart the particles are. Anton Zeilinger, having spent decades looking for ways to put quantum entanglement to use, showed in 1997 that it was possible to transfer information between particles, a process called 'quantum teleportation'. Along with another property of quantum particles called superposition, manipulating these became the basis of technologies such as quantum computing and quantum encryption, allowing quantum computers to solve problems that are deemed intractable with today's digital/classical computers.

While a classical computer uses binary bits that can only depict either a 0 or 1, quantum computers use qubits that can depict a 0, 1, or any combination thereof (superposition). Therefore, the computation capacity of a quantum computer grows exponentially since n number of qubits can represent 2n bits of information. Moreover, entanglement makes it possible for qubits to work together and represent multiple combinations of values simultaneously, allowing quantum computers to solve complex computational problems exponentially faster than classical computers.

Though quantum computers are in a relatively early stage of development, with engineers still experimenting with technologies to develop a full-scale quantum computer (e.g., superconducting, trapped ions, photonics, and silicon, among others), nevertheless, they are sufficiently advanced to explore applications and associated algorithms that address complex problems that are technically and economically impossible to solve using classical computers.

IBM led the way in 2016 with a 5-qubit computer and then a 20-qubit one in 2017. In 2019, Google was the first to demonstrate 'quantum supremacy', using a 53-qubit noisy intermediate scale quantum (NISQ) machine to perform in minutes a calculation that would have taken the world's fastest supercomputer more than 10,000 years. In 2021, a team of researchers from the University of Science and Technology of China used a quantum computer to simulate the behaviour of a molecule of hydrogen.

IBM, which has invested heavily in quantum computing, reckons it can build a 1,000-qubit machine soon. Both it and Google talk of a million qubits by the end of the decade.

QC, Climate, Cancer, and Space

Quantum Computers (QCs) can be used to model the climate more accurately, which would help us to understand and predict climate change and cope with one of humanity's biggest foreseeable challenges. QCs could even be used to model the behaviour of clouds, which play a major role in the Earth's climate. At least three major reports have elaborated on these possibilities.[1]

Cancer research is another area with fast-moving developments due to QC. IBM and the University of Chicago are using QC to simulate the behaviour of proteins involved in cancer, in particular, the protein p53, which is a tumour suppressor protein and responsible for repairing damaged deoxyribonucleic acid (DNA), but when mutated, it can lead to cancer. By simulating its behaviour, researchers hope to develop new drugs that can target it. Similarly, Google AI is using QC to identify new drug targets for cancer by scanning large databases of molecules to identify those that are likely to interact with specific proteins involved in cancer. This algorithm has been used to identify a number of new drug targets for cancer, including a protein called KRAS, which is involved in many different types of cancer. The Fraunhofer Society in Germany is using quantum computing to personalise cancer treatment.

In the arena of space exploration, quantum sensors have been developed to be used in space to study the Earth's atmosphere, the surface of other planets, and even the interior of stars. Quantum navigation is a new way of navigating in space that uses quantum sensors to measure the position and velocity of a spacecraft. Quantum communication cannot be intercepted and could be used to send confidential information between spacecraft or to communicate with astronauts on long-duration

---

[1] Quantum Computing for Climate Change: https://www.mckinsey.com/business-functions/mckinsey-digital/our-insights/quantum-computing-just-might-save-the-planet

How Quantum Computing Can Help Tackle Climate and Energy Challenges: https://eos.org/features/how-quantum-computing-can-tackle-climate-and-energy-challenges

Quantum Computing for Climate Change: A Research Agenda: https://arxiv.org/abs/2007.03032

space missions. Quantum computers are also being used to simulate the behaviour of materials at the atomic level. This could be used to design new materials with improved properties for use in space exploration, such as solar cells, batteries, and radiation shielding.

A team of researchers from the University of Vienna are using quantum computers to study the behaviour of gravity at the quantum level. They are using this research to understand how gravity works at the smallest scales. Another team of researchers from the University of Arizona are developing models of consciousness that incorporate quantum mechanics. They are using this research to understand the relationship between consciousness and the physical world.

The Interface Theory of Perception (ITP)

Donald Hoffman, a cognitive scientist who proposed a theory called the Interface Theory of Perception (ITP), has argued that spacetime is not fundamental but rather an adaptive illusion that our brains create in order to help us navigate the world.

Based on the idea that our brains are constantly making inferences about the world around us and that these inferences are based on our sensory experiences as well as influenced by our evolutionary history, he postulates that our brains have evolved to see objects as solid and three-dimensional, even though they are actually made up of atoms and molecules that are constantly in motion. According to the ITP, spacetime is one of these inferences. Our brains create the illusion of spacetime in order to help us make sense of the world around us. However, spacetime is not a real physical entity but simply a way for our brains to organise our sensory experiences.

Quantum computing could potentially be used to validate the ITP and find a way beyond our spacetime because quantum computers are able to simulate the behaviour of physical systems at the quantum level. This means that quantum computers could be used to simulate the way that our brains create the illusion of spacetime. If quantum computers are able to simulate the ITP, then this would provide strong evidence that spacetime is not fundamental. It would also suggest that there is a way to transcend spacetime. However, it is important to note that the ITP is still a controversial theory. There is no consensus yet among scientists about whether or not it is correct.

The Problem of Encryption—SNDL Attacks

Quantum technology's most dramatic impact is on the global cyber security architecture. Quantum computers will pose a threat to how we transmit and store confidential data like bank transactions, sensitive government information, and intellectual property. That is because, unlike existing computers, quantum computers will be able to crack the encryption systems that provide secure data communication and storage and underpin the global economy.

Since criminals and other adversaries know that this will be possible one day, they are collecting encrypted 'secure' sensitive data today, a phenomenon known as 'store now, decrypt later' (SNDL) attacks—stealing data for future decryption with quantum computers. According to America's Department of Homeland Security, the decryption of this data could be feasible as soon as 2030.

It is obvious that only data that will keep its value years from now will be targeted for decryption. SNDL attacks are being perpetrated right now, and once stolen, there is no way to protect encrypted data. Several countries are actively engaged in building quantum computing hardware for the sole purpose of defeating public-key cryptography and securing their own critical systems and data.

In response, researchers are developing quantum-resistant cryptographic schemes—new forms of encryption that even future quantum computers will be unable to crack. These are known collectively as 'post-quantum' cryptography (PQC).

New-age technologies like Bitcoin mining that rely upon solving puzzles with available computing resources will not work when the first QC starts participating in the network. The whole 'decentralised' financial system initiative will need major re-think and evolution once QC comes into play.

As in AI, China intends to lead the world in quantum technology with its national laboratory for quantum science, triggering Washington's efforts to create the National Quantum Initiative, which some observers have compared to America's nuclear programme of the 1940s. The European Union, too, has launched a quantum research initiative, and most recently, in April 2023, India launched its National Quantum Mission to join the global effort in a big way.

When 2+2 Equals 5 or 22

AI and quantum computing have a significant impact on each other. Quantum computing will enhance machine learning algorithms by providing faster computation for complex problems. Quantum algorithms, such as quantum support vector machines and quantum neural networks, will enable more efficient training and inference processes, leading to advancements in AI capabilities.

Optimisation is a fundamental aspect of many AI algorithms, where quantum computing can potentially offer improved optimisation techniques through algorithms like quantum annealing and quantum-inspired optimisation algorithms, leading to improved AI applications.

On the flip side, AI techniques, such as machine learning and data analysis, can assist in analysing and interpreting quantum experiments and simulations. AI algorithms can help identify patterns, optimise experiments, and extract insights from quantum data, thereby accelerating progress in quantum research. AI techniques can also inspire the development of algorithms that simulate quantum computations on classical computers. These quantum-inspired algorithms, such as quantum-inspired neural networks or quantum-inspired optimisation, attempt to mimic the behaviour of quantum systems to solve certain problems more efficiently.

To Sum Up

We are willy-nilly caught up in the revolution as we write (or speak), a revolution rapidly unfolding before us every day, redrawing the contours of everything we know. As tech takes over our lives, humans could do more on what they are meant to do, find their true purpose, if they ever find that, and become highly efficient in whatever they are doing. That's a best-case scenario. The worst case is that we do not adapt fast enough, leading to societal chaos. We could also see a mix of both, and not be on either extreme but in the middle.

Sam Altman recently spoke to the U.S. Congress about a scenario where AI becomes curious, leading to ambition, greed, and diplomacy, and somehow starts dominating the world. That's a doomsday scenario. There is no point worrying too much about it because the genie is already out of the bottle. We can't stop it anymore.

CHAPTER – 1

Humans and Transportation

Four Legs to Jet Packs

What is it about humans and transportation?

If getting from A to B were all it meant, our world would be different. However, acquiring a pair of 'sexy wheels' is about everything but that. Driving a Hummer to pick up groceries, flying business when the economy class gets you there in the same amount of time, or sailing across the world for kicks is not just about transportation but about us as individuals, too. There is the thrill, joy, and adrenalin, all of which must be tempered by the rights of others, traffic rules, and enforcement. Trainspotting is as serious a hobby as any but won't impress a plane spotter. Well, some of us are set to be spotting rockets and starships soon, no worries there.

When our ancestors became bipeds, walking was all they knew. As hunter-gatherers, they were perennially on the move. As they settled into farming and domesticated animals, maybe 10,000 years ago, the first 'means of transportation' arose. It is estimated that horses were domesticated around 6,000 years ago. If our ancestors were anything like us, you can bet that's when the competition, too, arose. Who owned a horse? Whose was faster or stronger? Black, brown, or white? Did the chief's horse have the tallest plume?

Ships came next, and we can date them up to 5,000 years ago. For the first powered flight, fortunately, we have the exact date – 17 December 1903 – when Orville Wright flew for 12 seconds over 36 metres. By then, in 1886, Carl Benz had already created a 'vehicle powered by a gas engine', and steam locomotives were hauling goods and passengers on iron tracks in the colonial world from India to the Americas.

From then till now, the transport industry has experienced a metamorphosis. We have hybrid and electric cars but talk of flying taxis, jetpacks, and drone delivery. Hyperloops and self-driving cars are par for the course. We travel higher, faster, and over longer distances than it has even been possible and don't bat an eyelid. Captain Kirk wasn't wrong when he said 'Space – the final frontier'. We are getting there. In Style. Or are we?

Gridlocked in Style

In some of India's best-planned cities like Chandigarh and Bengaluru, don't be surprised if you can't find a parking slot in the area you visit. Or that the top speed that your Ferrari attains on the road is 20 kmph. As *The Economist* said in an article more than two decades ago, 'the point at which cars became secondary to the overall road system was reached in urbanised society some time ago. It matters less and less how fast your car goes and more how much the road system allows you to go.'

The commute is the leitmotif of urban existence, and words like rush hour, roadblock, diversions, and road rage are everyday vocabulary. For those who are more familiar with the

North, you will find that the police barrier is always designed to choke the last breath out of a gasping traffic flow. If nothing else, trust a VIP movement to mess up traffic on an otherwise calm day. Or, you may be a suburban or metro train commuter, carried in and out of trains on your regular stops as the flow of humanity rises and ebbs like the tides, hanging on to your bags and laptop for dear life.

Not that flying makes it any better. The skies are chock-a-block, too. In January 2023, thousands of flights had to be grounded due to the failure of a key pilot notification system called Notice to Air Missions (NOTAM) due to some unintentionally deleted files by a Federal Aviation Administration (FAA) operative. Hundreds of thousands of passengers were backlogged at airports across the USA. Airports are crowded, and systems groan and creak to cope with the rush of people trying to get somewhere else. Just ask a passenger transiting through Heathrow in London.

If the future were only an extrapolation of the present, at the pace at which we are adding cars, planes, trains, ships, and now drones, we would be staring at a gridlocked planet. Fortunately, that need not be the case if our planners figure it right. Unfortunately, we might not have such planners, or we wouldn't be in this mess in the first place.

Gaia's Fever

The gridlock, whether we solve it or not, poses a different challenge as well—the consumption of fossil fuels and its impact on climate change and global warming. Without delving into the hows and whys of this by-now well-understood issue, it is evident that future transportation cannot be driven by fossil fuels. That statement immediately morphs the problem from transportation and technology to economics, geopolitics, and warfare. Yes, warfare. Though we deal with Humans and Warfare in a separate chapter towards the end of this book, it is important to understand how it affects transportation.

An often-neglected aspect of transportation is its military application. The combined strength of the world's militaries translates into thousands and thousands of fighter and transport aircraft and helicopters, battleships and submarines, trucks, tanks, and whatnot. They are designed for many purposes, but fuel efficiency is not one of them. They guzzle fossil fuel. What's more, while there is enough academic research and statistical analysis of the (civilian) world's transportation needs and energy consumption, there is hardly any data in the public domain when it comes to the military. Each country is super secretive. How much wars, training exercises, and routine patrolling feed Gaia's fever is anyone's guess. Billions of barrels are extracted and stashed away in strategic oil reserves only to be converted into smoke.

Hence, if the future of transportation is clean energy, how would it include the military in its ambit with all the overt and covert warfare that carries on 24x7x365? Are we looking at electric tanks and fighter jets? I think not. However, autonomous tanks and fighters may soon be a reality, like the autonomous cars on our streets. (Or maybe they already are—remember, this is the military.)

In reality, that last sentence deserves to be turned on its head. Historically, it is military innovation that has driven innovations in transportation technology. The world wars birthed the aviation industry by producing aircraft at an industrial scale, likewise for shipping. Some of the biggest manufacturers in the transportation arena owe their origin as wartime suppliers—Boeing, Ford, General Motors, Fiat, Mitsubishi, and many others. Indian conglomerates like the Tatas have also been military transport suppliers for decades. Transportation technology and the military are tied in a Gordian knot, which would need to be unravelled before Gaia can be healed. To say that we need green fuel and clean energy to deliver bombs and missiles to obliterate fellow humans would be a sad statement to make, yet that would be the implication if we were to clean up the planet.

When the Chips are Down

A corollary to the above has recently been witnessed by the world. Semiconductor shortages are nothing new. In June 1983, the *New York Times* reported a semiconductor shortage due to a boom in personal computers that could 'last well into the Orwellian year'. Customers had to wait 20 weeks for the delivery of a computer. Since then, there have been periodic booms and busts in the supply chain, affecting many IT-driven industries but not (until recently) transportation.

Things took a turn with the supply chain shocks delivered by the Covid epidemic, specifically for the car industry, because by then, the industry had undergone what *The Economist* calls 'Teslafication'. Automobiles were now 'computers on wheels, running as much on processing power as the horse variety', and the global chip shortage could claim its newest victim.

Elon Musk's, and hence Tesla's, philosophy of vertical integration—right from designing their chips to mining their nickel—represents the opposite end of the spectrum of where other car manufacturers were positioned with their just-in-time outsourced supply chains. The pandemic wreaked havoc on them. Companies like Toyota reported production cuts as deep as 40 per cent on occasion, and car buyers waited months to buy what used to be all in a day's work.

Similar could be the case with electric car batteries and software upgrades. Your car would now require another kind of boot as well (pun intended)!

Supply chain linkages are directly affected by the prevailing geopolitical tensions and the winds of globalisation and deglobalisation as they keep changing direction, the latest disruptions emanating from the Ukrainian-Russian conflict. The fragility of our future technology-driven transportation systems would be a key concern as we peer into our crystal ball.

Traveller's Techtopia

To understand the utopian transport systems we could be headed towards, there is an interesting convergence of technologies from different domains, bringing innovative solutions, some of which will succeed even as others fail.

Take the algorithmic problem of efficiently scheduling and routing cabs, which Uber and Lyft solved and dramatically changed the way people in large cities find transportation. With an app, riders could summon a vehicle to their location any time they wanted. Combine this with research in alternative fuels, electric mobility, lightweight composites, drone technology, autonomous vehicles, and, of course, AI, and you are looking at a landscape from a sci-fi film. Here are some glimpses.

Lightweight Vehicle Materials

Various companies today are utilising alternative materials to produce lightweight vehicles. For instance, Ford is using natural fibre-reinforced composites in the armrests of their Lincoln MKX SUV, while 3M has developed hollow glass microspheres that reduce composite weight by up to 40 per cent. Tata Steel has launched Coretinium, an optimised polypropylene honeycomb core used to lighten the weight of floors and sidewalls in buses and trailers.

Composite materials like aluminium, magnesium, and steel blends offer better energy absorption than solid metals like steel alone, providing more protection in the event of an impact. Lighter cars require less energy to accelerate than heavier ones, making lightweight materials more beneficial for both automakers and consumers.

Drone Buses and Autonomous Helicopters

Drone buses have arrived. These can ferry over 40 people, fly fast, and, unlike aeroplanes, can land and take off closer to city centres. This announcement has been made by the New York-based start-up Kelekona. It has gone on to say that, by 2024, it will have the first drone bus ready for its maiden passenger flight. The company says that these buses can be used for cargo as well, but its main aim is to revolutionise public transport. To be built of 3D-printed composite and aluminium, a 30-minute flight from Manhattan to the Hamptons is likely to cost $85. Uber and several Chinese businesses are also investing in this developing transportation technology.

In 2017, Aurora Flight Sciences, a company that focuses on aviation research, demonstrated a completely autonomous military helicopter flight. Years later, Airbus has now accomplished a test flight of Vahana, a self-piloted, all-electric, single-seat, and tilt-wing electric vertical take-off and landing (eVTOL) demonstrator. Early last year, the Pentagon's Defense Advanced Research Projects Agency also successfully tested a helicopter equipped with experimental autonomous flight software. An autonomous ambulance has already been tested in Israel. In 2016, the AirMule, which can carry two people up to a total of 450 kg, was designed to take off and land in places where a traditional helicopter could not.

The Hyperloop

In 2012, Elon Musk unveiled the Hyperloop, a high-speed transportation concept that aimed to combine the affordability of road transport with the speed of air travel. Working in partnership with Tesla and SpaceX, Musk's Hyperloop operates using reduced-pressure tubes containing pressurised capsules and promises to transport passengers from San Francisco to Los Angeles in just 35 minutes, travelling at speeds of up to 700 miles per hour.

Virgin Hyperloop, headed by Richard Branson, successfully created a working prototype of the Hyperloop system in 2017 and has since accomplished a fully operational system. One of the most significant advantages of the Hyperloop system is its ability to offer an on-demand service for travellers, with capsules departing every few minutes, reducing waiting times compared to other high-speed rail networks. Musk's decision to open-source the concept has also spurred the creation of new businesses and groups dedicated to advancing this innovative technology. As Virgin Hyperloop One continues to work with Hyperloop Transportation Technologies, the dream of bringing the Hyperloop to life by 2030 is becoming closer to reality.

Hypersonic Air Travel

In November 2003, the Concorde Supersonic airliner flew its final flight, surpassing the speed of sound at Mach 1 or 770 mph. However, there is a new wave of aviation companies looking to advance technology and fly passengers at hypersonic speeds, potentially reaching Mach 5 or 3,800 mph, shortly. With such speeds, transatlantic travel from New York City to London could be reduced to just two hours. Reaction Engines, a British company, has made progress in testing an engine cooling system that could potentially support aircraft flying at Mach 3.3 to achieve Mach 5.

Maglev Trains and Decarbonising Railways

Despite sounding like a futuristic concept, maglev technology has been implemented into transport networks across the world since 1984, with trains running on maglev systems. In 2015, a Japanese maglev train achieved a speed of over 600 kmph, while researchers at China's Southwest Jiaotong University are testing an ultra-fast bullet train prototype based on maglev technology that has the potential to reach speeds of up to 1,000 kmph. Super-maglev trains, which use a vacuum tube to reduce air friction and allow for increased speeds, represent the next generation of maglev technology.

The railway industry has found various applications for AI technology, including asset management, predictive maintenance, and emergency notification. Neural networks and deep learning algorithms are used to optimise train scheduling, reduce delays, and improve transportation services. Governments are seeking to decarbonise the rail industry despite railways being the most sustainable form of transportation and following ambitious net-zero emissions plans. The prevalent decarbonisation solutions include the adoption of battery technology, hydrogen fuel cells, or electric trains to replace diesel trains. To further reduce carbon dioxide emissions, rail operators are turning to electric locomotives that harness energy from renewable sources, such as solar or wind.

Roads of the Future

Advancement in transportation does not only mean modes of transport but also the surfaces that they traverse. Examples are elevated cycle paths and smart roads. The first such elevated cycle path has been constructed in the southeast Chinese city of Xiamen. At 17 feet above the road, this 7.6 km track, with a capacity for over 2000 bikes, has connections to public modes of transport at 13 points. Copenhagen, a bike-friendly city that already has a cycle super-highway that connects the city with Albertslund, which is 22 km away, plans to build a network of 28 cycle superhighways, which will cover 500 km. London and New Delhi have unveiled similar plans.

Smart roads, having IoT-enabled road sensors to instantly communicate with smart cars about hazards and adverse road conditions, are here, too. The European Union (EU) is funding cutting-edge transportation technologies to enable wireless communication between smart cars and infrastructure nodes. Sweden is trying to transform road stretches into electrified tracks that will recharge electric vehicles like cars and trucks while they are in motion. In the future, wireless battery chargers will be installed beneath roads.

Elon Musk is revolutionising conventional road design with his groundbreaking concept of underground roadways. In his visionary approach to addressing urban infrastructure challenges, roads are made three-dimensional, where cars are taken underground instead of flying in the sky, offering better weather resistance and reducing traffic congestion.

Futuristic Elevators

If you thought that the only thing that elevators did was to move you up and down, then read on. Humans have never stopped inventing and innovating, not even when it comes to elevators! ThyssenKrupp, a German elevator manufacturer, is developing a revolutionary elevator system that is rope-less and moves sideways. Unlike traditional elevators that move up and down a vertical shaft, this new system consists of multiple cabins that run on an electromagnetic track. Using magnetic levitation, the cabins move up one shaft, travel horizontally, and descend through another shaft, creating a continuous loop. Carbon-fibre bearings, called slings, are mounted on sophisticated switches that guide the cabins and allow them to change direction. ThyssenKrupp has also recently introduced MAX, the elevator industry's first real-time, cloud-based, predictive maintenance solution, which utilises AI and ML to predict issues before they arise, improve reliability, and flag component replacements.

Futuristic Fuels

Although petroleum and diesel still dominate fuels, there is an increased viability of alternative energies. Sustainable transportation technologies such as electric vehicles and natural gas are advancing, and the industry is optimistic about diversifying into fuel alternatives for the future.

Biofuels such as ethanol, methanol, biodiesel, bio-crude, and methane are being produced from biomass as a promising alternative to traditional transportation fuels. Currently, ethanol stands as the foremost biofuel in widespread use. Methanol is another alcohol that can serve as a transportation fuel. Biodiesel is a renewable diesel fuel substitute that can be made by chemically combining any natural oil or fat with an alcohol, typically methanol.

Hydrogen is showing a lot of promise as a clean alternative fuel that can power fuel cell vehicles with zero emissions. Fuel cell electric vehicles (FCEVs) rely on fuel cells to combine pure hydrogen and oxygen, producing electricity to power the vehicle's battery. The adoption of hydrogen fuel cells, though, faces challenges, particularly in terms of cost. The main advantage of FCEVs over electric options is their range, with hydrogen fuel cell trucks able to travel over 1,000 miles at a time, making them ideal for long hauls. Despite its clean emission profile, the sustainability of hydrogen production is a concern, as some methods rely on fossil fuels and emit greenhouse gases.

Natural gas has cost advantages over gasoline and diesel fuels and, hence, can be used as an alternative and natural form of energy. Compressed natural gas (CNG) is used in consumer vehicles and heavy-duty transportation, offering a low and stable price point and reduced emissions compared to diesel. Liquefied natural gas (LNG) is less popular because it is costly to produce, store, and transport, emitting higher levels of greenhouse gases through its cooling process. Renewable natural gas (RNG) is chemically the same as conventional natural gas but made from renewable source materials like animal waste, landfill gases, and decaying organic matter, which eliminates carbon emissions in some cases. California is one city that is leading the charge in the utilisation of RNG for transportation fuel, with regulations and incentives to make it more cost-competitive with diesel fuel.

Next-Generation Lithium-Ion Batteries

Electric-powered or hybrid vehicles have received significant investments from major automobile manufacturers. Despite higher costs and limited range, consumers continue to buy them due to their cost-effectiveness and positive impact on the environment. However, to help electric cars achieve significant market share, the limitations of lithium-ion batteries are being addressed. Researchers are exploring ways to enhance battery structure and electrode chemistry to increase their capacity to hold a charge for extended periods. Silicon nanoparticles represent a promising alternative to the conventional carbon graphite electrode, capable of boosting energy density by up to 20 per cent, thereby facilitating longer-lasting batteries.

Teleportation

The concept of teleportation has long been a staple in science fiction, captivating the imaginations of people around the world. While teleportation remains purely fictional in the present, scientists have explored the idea and made progress in the field of quantum teleportation, which involves the transfer of quantum states between particles.

Quantum teleportation enables the transfer of quantum information between particles, such as photons or atoms, over large distances. This process relies on the principle of entanglement, where two particles become intrinsically linked in such a way that the state of one particle is instantaneously connected to the other, regardless of the physical distance between them (the same phenomenon that you read in our opening chapter on AI), and effectively transfers the quantum state from the sender's particle to the receiver's particle.

While quantum teleportation holds promise for applications in quantum computing and quantum communication, it is not directly applicable to the teleportation of macroscopic objects or people. The complex systems and intricate quantum interactions required for teleporting large-scale objects pose significant challenges that are currently insurmountable based on our current scientific understanding and technological capabilities.

Space Travel

Laika, a terrier mutt from the streets of Moscow, was just two years old when she became the first Earthling to orbit our planet. Her historic launch occurred only a week after she was chosen. Laika's flight in 1957 helped pave the way for space exploration.

The journey to space initially began as a competition between the US and the former Soviet Union, both vying for power and global dominance. Yuri A. Gagarin became the first human to venture into space. Having conquered Earth's orbit, the focus shifted towards the moon. In 1969, the US successfully landed two astronauts on the moon's desolate surface, followed by five more manned missions. The National Aeronautics and Space Administration (NASA) then launched space probes to explore the solar system, and manned space stations sparkled in the sky. The development of reusable spacecraft, such as the space shuttle orbiters, made it possible to transport astronauts and satellites into orbit. Remarkably, space travel technology advanced significantly in just three decades.

The adventure has already begun. Planetary Resources plans to send robotic astronauts to the asteroid belt to extract valuable metals. SpaceX recently reached a significant milestone by launching astronauts from the United States to the International Space Station. The relationship between NASA and private companies is being enhanced through the expansion of the Commercial Crew Program. NASA has entrusted SpaceX and Boeing with the responsibility of developing spacecraft that can transport humans into orbit. After the vehicles have been constructed, the companies will maintain ownership and control over them, enabling NASA to send astronauts into space at a significantly lower cost than using Russia's Soyuz spacecraft. In May 2020, NASA astronauts Doug Hurley and Bob Behnken made history by being the first crew to launch from the United States in almost ten years onboard SpaceX's Crew Dragon spacecraft. Blue Origin and Virgin Galactic are focused on sub-orbital space tourism, offering customers a unique experience. A test launch video from Blue Origin's New Shepard displayed stunning views of Earth from the cabin, and the journey was relatively peaceful for its first passenger, a test dummy named 'Mannequin Skywalker'. Virgin Galactic is currently conducting test flights of its suborbital spaceplane, which will provide paying customers with approximately six minutes of weightlessness as it travels through Earth's atmosphere.

In *The Hitchhikers Guide to the Galaxy,* there comes a time when the Earth is demolished to make way for a galactic freeway. Well, we won't let Gaia be demolished, but building galactic freeways may yet be our destiny. When the mountains and the seas seemed insurmountable, we triumphed. Our dreams of intergalactic travel will also turn real, all in good time.

CHAPTER – 2

Humans at Work

Will AI Gobble Jobs a.k.a. The Great Churning

Fear, uncertainty, and doubt is commonly known in military parlance as the FUD factor. If there is one way in which AI has impacted our workspaces, it has been through FUD. Workers are scared out of their wits that their jobs will be taken by AI, machines, robots, or a handful of supersmart humans. Terrifying predictions abound, and Tyna Eloundou of OpenAI and colleagues have estimated that 'around 80 per cent of the US workforce could have at least 10 per cent of their work tasks affected by the introduction of LLMs'.[1]

Sure, the mundane jobs will go, where you're doing one thing to another, whether researching something, posting content for marketing, or driving a car—self-driving cars have been experimented with for the longest period over the last few years if not decades. AI can genuinely beat their last challenge,

[1] *Economist* dated 7 May 2023

the last mile, and help build driverless cars, so the drivers' jobs would be in question.

The IT and tech services jobs—think IBM, TCS, Infosys—involve coding business logic. Given a project, if there are ten people on an IT team, and the client explains the business logic that he needs in English, then they code it up in some programming language like C++, Python, etc. Well, AI does this much better. Two things can happen—you can take those ten people and start doing ten times the work. So, you can pick up ten more projects with the same people and drive down the cost. That is an impact of any big technology—disruptive. Or, for the same project, you may need only one person who coordinates with AI and gets it done. Either the TCS client projects have to grow 5x or 10x to maintain their revenue growth, or the cost will drop five to ten times over time within a decade. They start cutting down, and AI starts doing everything—it is basic economics.

Think about doctors. If you go to a doctor, they are so busy. Their training and education take long, sometimes up to ten years—MBBS and other advanced degrees—and they start by asking you, 'Okay, what seems to be the problem?' You say you have these symptoms, and this is going on. They reply, 'Okay, try this medicine and come back in a few days if it doesn't work.' An AI can solve this much better. Knowledge, in any knowledge economy, can be processed much more efficiently by AI. It already has access to all of it, and it's just a matter of time before humans start using AI to do more, and the cost drops dramatically, or humans start doing less, and machines start doing more. There could be a combination of both, and job losses may happen in between. People must think of smarter ways to use AI rather than for routine work. There would be more AI managerial jobs, I imagine.

In the short term (say within 5–10 years), there could be many job losses, productivity increases, or a combination of both, and the next few years promise to be very interesting. Research by Goldman Sachs, a bank, suggests that 'widespread AI adoption could eventually drive a 7 per cent or almost $7 trillion increase in annual global gross domestic product (GDP) over a ten-year period.' Runaway growth is not impossible, as some economists are predicting.

AI Taxes?

Governments are unlikely to be ahead in any of this. They may not be able to regulate AI technology well enough because the pace of development will be far ahead of its application and understanding. Of course, governments can and will tax anything and everything, and I suspect that an AI tax surcharge could well be on the way. I'll take an example to make it real for you. Today, say, Uber charges ₹15 a kilometre for a ride, and ₹5 go to Uber, ₹5 go for car maintenance and petrol, and ₹5 go to the driver. If AI replaces the driver and lowers the maintenance cost, Uber may save an additional six to seven rupees out of that ₹15. Of that, ₹2 may go as an AI surcharge tax to the government, ₹2 as a discount to the consumer, and ₹2 more to Uber for implementing the technology, making the process more efficient and more sustainable. And then, at some stage, the governments would have to start considering universal basic income because how would humans be paid if machines were to do everything?

The Vulnerable Sectors

Call centres or support teams, IT services, driving jobs, and many other mundane and simpler jobs involving logic processing or data conversion will likely start disappearing at an exponential pace. People working those jobs will have to reskill or step up and justify how they fit the newly evolved job profiles to survive in the market. Looking back, humans have always adapted to changes and found a way, and this time, too, one would hope it happens. However, the short-term chaos due to the pace of change and the lack of time offered to humans to adjust is understandable.

The Not-So-Vulnerable Sectors

The area likely to be least impacted by AI is blue-collar work, which today accounts for about 20 per cent of the GDP of developed countries in the West. Even with an AI diagnosis, you would still need a plumber to fix a leak. For a long-term bedridden patient, you will need a nurse to physically help out.

Other industries where human-to-human contact is an essential part of the service, such as hospitality and medical care, would also be less vulnerable.

The New Job Market

The most compelling pro-AI argument for the economy is that it is likely to create new occupations which cannot even be imagined at present. This trend has been observed in economic history at every major technological inflexion point, be it the Industrial Revolution, the invention of electricity or the printing press, the coming of railways, then aviation, and the dawn of the Information Age, followed by the coming of the Internet. With each invention or innovation, jobs have been lost, and newer ones have been created in larger numbers.

Could AI Lower Human Productivity?

Technology may not always be a boon, leading to increases in productivity. The best example to explain this is the proliferation of smartphones. While they allow instant communication, they can be a severe distraction as well. The ease of 24/7 email and work communication is now facing a backlash from employees everywhere as it disturbs their work-life equilibrium and generates mental stress.

Similar has been the case with the technology-induced work-from-home (WFH) boon that came partly out of the compulsions of Covid and, later, out of the desires of employees to continue in the same mode. Initially, employers adopted it under compelling circumstances and were pleasantly surprised by its effectiveness and results. It gave rise to visions of a borderless world with employees working from beaches and mountains, happy and highly productive, while businesses saved on office overheads, and cities got rid of commuting and environmental problems. Alas, this wasn't to be. While the underlying technology has only improved and matured to support such an ecosystem, humans gradually wore out the euphoria. Productivity from homes plummeted, workcations could only be stretched so much, and reality caught up. So, while the employers now want their staff back in offices, the latter are resisting. A kind of quasi-hybrid model is now being tried at a lot of workplaces, and the final word is yet to be said on the issue.

The workplace is also about the physical bonding of employees and teams, the exchange of ideas and knowledge, and the cultural and psychological aspects of corporate work, and therefore, productivity is clearly not a function of technology alone. Hence, while AI may be great tech, will it make humans more or less productive? It needs to be debated.

> Forget 9-to-5. The future of work is 24/7... seconds, that is, between when you think of an idea and a machine does it for you.
>
> (Credit: ChatGPT)

CHAPTER – 3

Humans at Play

Gaming and Convergence

All work and no play makes Jack a dull boy.

If current trends are any indication, the future Jack is poised never to have a dull moment—for 'gaming' is already big and growing bigger. Some 3.2 billion people played video games in 2022, or about four in ten people worldwide.[1] In rich countries, two-thirds of people play, nearly half of them women, with gaming no longer the domain of only young people.

This massive demand has helped gaming companies grow exponentially such that seven of the ten most valuable tech firms are active in gaming, including Apple and Google, whose over 60 per cent of app store sales now come from games. Disparate players like the e-commerce giant Amazon and the chipmaker Nvidia are also streaming games to get their foot in

[1] https://www.economist.com/special-report/2023/03/20/ready-player-four-billion-the-rise-of-video-games

the door, while China's Tencent has made gaming central to its business model. Meta's Metaverse attempt is nothing but a pivot from social networking to virtual reality, with gaming as the main use case.

People paying to watch others play video games is the most unexpected market that has grown massively in the last decade or so. Esports has now been included even in the Olympics.

The bigger story in gaming is the convergence of other forms of entertainment with it. It has expanded its reach to encompass other forms of entertainment, with over-the-top (OTT) streaming services and apps embracing gamification to enhance user experiences.

Companies like Netflix and Amazon Prime Video have recognised the immense popularity of gaming and the potential it holds for enhancing their offerings. As a result, they have ventured into gaming to diversify their content libraries and engage users in new and exciting ways. OTT streaming services have started incorporating interactive gaming experiences, allowing users to make choices that shape the narrative in choose-your-own-adventure style stories. These games blur the lines between traditional streaming and gaming, immersing users in a more interactive and participatory form of entertainment. By combining story-telling with interactivity, OTT services create a unique and personalised experience for their subscribers.

Beyond streaming services, gamification has found its way into a wide range of OTT apps, transforming various aspects of our daily lives into engaging experiences. For example, fitness apps utilise gamification elements to motivate users to achieve their health goals. By turning exercise routines into interactive challenges, users are incentivised to stay active and track their progress. Leaderboards, achievements, and rewards foster a sense of competition and accomplishment, making fitness more enjoyable and immersive.

Virtual concerts in games are already a reality. In 2019, Marshmello, a world-renowned DJ, made history by holding a live concert in the game Fortnite, and it was attended by over 10 million people. Since then, other famous musicians such as Travis Scott and Ariana Grande have also had their in-game concerts.

Language learning apps employ gamification techniques to make the process of acquiring a new language engaging and entertaining. Through gamified quizzes, challenges, and virtual rewards, users are encouraged to continue learning and improving their language skills. By adding a playful and competitive element to language learning, these apps enhance user engagement and create a more effective learning environment.

Gaming is also influencing core sectors like education and learning. Educational games and gamified learning platforms have gained popularity as they offer interactive and engaging experiences that promote active participation and knowledge retention. From language learning to math and science simulations, gamification enhances the learning process by making it fun, immersive, and personalised.

In healthcare, gamified applications and devices are used to motivate patients and encourage them to adhere to treatment plans, monitor their health, and engage in physical therapy. VR and AR technologies are also employed to create immersive experiences for pain management, phobia treatment, and surgical simulations.

Gaming has always been an integral part of the corporate world's arsenal. In marketing and advertising strategies, brands incorporate gaming elements into their campaigns to create interactive and memorable experiences for consumers. Gamified advertisements, interactive quizzes, and reward-based loyalty programs are just a few examples of how to drive customer engagement. In corporate training and team-building activities, serious games, simulations, and virtual environments are employed to enhance employee skills, foster collaboration, and promote problem-solving. By gamifying training programs, companies can create engaging and interactive learning experiences that encourage active participation, improve knowledge retention, and provide real-time feedback.

Gaming has also found applications in promoting social impact and driving behavioural change. Serious games tackle important societal issues such as sustainability, public health, and civic engagement. They provide a platform for raising awareness, educating users, and inspiring action. By combining entertainment with meaningful content, serious games can influence attitudes, behaviours, and decision-making processes.

Social Media and the Metaverse

The roots of social media stretch back to the 70s with the advent of the Internet. The earliest iterations were far from the platforms we know today but set the stage for online interactions. Bulletin Board Systems (BBS) and Internet Relay Chats (IRC) allowed users to share messages and files and engage in real-time discussions. However, these services were mainly text-based and lacked the interactive features associated with modern social media.

The first recognisable social media site, SixDegrees, was launched in 1997. It allowed users to create profiles, make friends, and post bulletins. Fast forward to 2002, and Friendster appeared on the scene, setting the stage for the social media era as we know it by adding a new degree of interactivity.

MySpace, launched in 2003, was the platform that made social networking mainstream. It allowed for greater personalisation, including the ability to customise profiles with backgrounds, music, and other multimedia. Before Facebook gained dominance in India, Orkut, owned by Google, was the most popular social media platform in the country. Launched in 2004, Orkut quickly gained popularity in India and Brazil. Named after its creator, Google employee Orkut Büyükkökten, the platform allowed users to create profiles, add friends, and join communities of interest.

Then came the giant leap—the launch of Facebook in 2004. Facebook brought a level of scalability and usability that hadn't been seen before. It quickly became the leading platform, with Twitter following closely behind in 2006, introducing the concept of microblogging.

We have already seen how Facebook's (Meta's) idea of the social future of humans is embedded in the Metaverse. Whether the Metaverse evolves the way Meta conceives it or not, what is undeniable is that AI is already shaping social media through recommendation algorithms and targeted ads. In the future, it will hyper-personalise user experience by tailoring content to individual preferences, creating a more engaging and relevant social media environment.

Blockchain technology may possibly lead to decentralised social media platforms, giving users more control over their data and privacy. This approach could shift power away from central corporations and back into the hands of the users, thus promoting transparency and data security.

As social media's impact on society becomes more apparent, we can expect increased scrutiny and regulation. Policies on data privacy, misinformation, and online harassment will likely become more robust, shaping how platforms operate. Social media, arguably, has already led to ample time wastage for humans, irreversibly, for decades now.

AI could play a vital role in moderating content on social media, identifying and removing inappropriate content, hate speech, fake news, and online harassment, and creating safer online spaces.

On the flip side, AI's ability to create highly realistic deepfakes or synthetic media would have implications for social media. While this technology could be used for creative and entertaining purposes, it also carries potential risks, such as the spread of misinformation or fake news.

Internet-of-Things (IoT)

With IoT making waves in various sectors, including healthcare, transportation, and manufacturing, one area that often gets overlooked is how IoT will reshape human leisure time in the coming decades. By connecting everyday objects to the Internet, IoT technology is set to redefine our recreational activities, hobbies, and downtime.

IoT is rapidly transforming our homes into smart homes, enabling appliances and systems to communicate with each other. This interconnectedness can greatly enhance our leisure time. If you enjoy cooking as a form of relaxation, a smart kitchen can suggest recipes based on what's in the fridge and start preheating your oven while you're still on the way home.

In outdoor recreation, IoT can enhance leisure activities. Connected devices such as fitness trackers and smartwatches would become even more sophisticated, offering personalised training recommendations or immersive AR experiences during hikes or city tours.

IoT has significant potential in the realm of travel and tourism. Smart luggage can be tracked to prevent losses, wearable tech can provide real-time translations, and smart hotel rooms can adjust to personal preferences for lighting, temperature, and entertainment. As more cities become 'smart cities', tourists will be able to interact with their surroundings in a more connected, immersive way.

Finally, as our world becomes increasingly connected, IoT might play a role in promoting the art of doing nothing—a luxury in our busy lives. Smart homes can take care of mundane tasks, freeing up our time to simply relax and unwind.

Live Sports

Sports enthusiasts often say that attending a live game is more than just watching it; it's an immersive experience. The future of live sports promises to be an exciting confluence of technology, fan engagement, and on-field action. The Metaverse could revolutionise sports entertainment by allowing fans to be a part of the action. By personalising their avatars with team jerseys, cheering on their favourite players, and celebrating with friends after the game, fans can experience a complete stadium atmosphere rather than simply watching the game on TV.

AI is poised to change the game in sports analytics and coaching. It can process vast amounts of data—player stats, game strategies, historical performance under specific conditions—to generate valuable insights. This can help teams prepare better game plans, spot patterns in opponents' strategies, and manage player fitness levels. Moreover, AI can be used to develop personalised training programs, identifying areas of improvement for each athlete and suggesting corrective measures. This technology can also help prevent injuries by predicting when players might be at risk based on their training loads and historical data.

IoT—with its web of interconnected devices—could significantly enhance performance tracking and safety in sports. Wearable IoT devices would monitor athletes' vital signs, hydration levels, fatigue, and stress in real time. This would provide invaluable data to manage players' health and optimise their performance. Moreover, IoT technology can also improve safety in sports. For instance, smart helmets in football or motor racing can monitor impacts and detect signs of concussion immediately, allowing for swift medical intervention.

Gamification—the application of game elements in non-game contexts—is another trend shaping the future of live sports. Fans might participate in fantasy leagues, predict game outcomes, or engage in e-sports, blurring the lines between being spectators and participants. Live sports events could also incorporate augmented reality games, where fans engage in interactive contests or challenges during game breaks. This would lead to higher fan engagement and a more immersive sports experience.

Tourism

Tourism, like many other industries, is being transformed by the rapid pace of technological advancements. With AI, the Metaverse, IoT, and gamification at the forefront, the future of travel and tourism will likely be an interconnected blend of physical and digital experiences.

Like every other sphere, AI will make travel experiences more personalised and convenient with suggestions based on travellers' preferences, past trips, and real-time data such as weather and local events. Chatbots and virtual assistants would provide instant responses to queries, streamline bookings, and offer 24/7 customer service in multiple languages, making travel more accessible and hassle-free.

Metaverse, too, offers exciting possibilities for the tourism industry with virtual tours of destinations, allowing travellers to explore places before they visit or experience inaccessible or remote locations. Imagine walking around the pyramids of Egypt or exploring underwater coral reefs from the comfort of your living room. Furthermore, it can serve as a platform for preserving historical and cultural sites. By creating virtual replicas of these sites, we would ensure that they're accessible to future generations and aren't lost due to natural disasters or human actions.

IoT would enhance travellers' experiences by creating a connected, seamless travel journey. Smart luggage trackers would help prevent lost baggage. IoT-enabled hotels would personalise guest experiences by adjusting room temperature, lighting, or entertainment options based on guest preferences. Wearable IoT devices could provide real-time, location-based recommendations, translate foreign languages, or make contactless payments, making travel more convenient and immersive.

And finally, gamification would make travel experiences more engaging and fun. Travel apps would incorporate games or challenges related to local culture or landmarks, encouraging tourists to explore and learn more about their destinations. Augmented reality games would transform physical locations into interactive experiences, adding a new dimension to sightseeing.

The Creator Economy

Traditional and mainstream forms of media and entertainment would continue to decline as personalised, niche, and community-driven media, known as the creator economy, gain popularity instead. Short-form videos and creator-driven platforms would remain more popular, particularly those that allow audiences to form personal relationships with their favourite creators and influencers. The Patreon platform saw a doubling of money paid to creators from $1 billion in 2020 to $2 billion in 2021, indicating a growing comfort among audiences to bypass networks and platforms to directly connect with their favourites. This trend is expected to accelerate in the future.

Throughout history, entertainment technology has continuously evolved and improved, from ancient tools such as the Hero of Alexandria and vending machines used to automate aspects of theatre performance to the AI-driven personalised recommendations of today's media and entertainment industry. As technology continues to advance, there is no stopping the evolution of the entertainment industry.

CHAPTER – 4

Humans and Knowledge

Epistemology

What do we collectively know, and how do we know what we collectively know? Well, that's epistemology for you—the branch of knowledge that studies knowledge itself. Very meta!

Knowledge has different aspects—its nature, extent, truthiness, etc. It exists in one's mind like a belief. Not every belief is true, so one's knowledge constitutes one's true beliefs (we are presupposing the existence of objective truths). Knowledge also requires justification—the right way of arriving at the true belief, i.e., sound reasoning and solid evidence.

Justification does not imply certainty. Humans can be wrong, and sometimes one's true beliefs may turn out to be false. To go any deeper, one would have to begin with an Epistemology 101 course, which, of course, is not the objective here.

The Historical Evolution of Knowledge

The evolution of human knowledge certainly did not begin with epistemology for pre-historic humans. With no alphabet, the most basic of tools, and as hunter-gatherers, they were in no position to theorise anything, least of all knowledge itself. Yet, it seems fairly obvious that they would have formed many 'beliefs', and they would have quickly established their truthiness, such as that sabre-toothed tigers don't make for good pets, and they would have fairly disseminated this 'knowledge' to, say, their children, who were yet to encounter a sabre-toothed tiger.

Or it could have been knowledge about the best hunting grounds, which berries were safe to eat, where water could be found, or even the patterns of stars in the sky.

So, we began acquiring, processing, disseminating, and evolving knowledge from the time we appeared on Earth. The first major inflexion point in this evolution was the invention of writing, which occurred in ancient Mesopotamia around 3200 BCE. The Sumerians recorded legal, economic, and religious texts, creating a means to transmit information through generations. The other ancient civilisations were hardly likely to be left behind, and the Ancient Egyptians contributed to geometry and medicine. The Rhind Mathematical Papyrus and the Edwin Smith Papyrus are testament to their ingenuity in these fields. The construction of the pyramids showcased not only their architectural prowess but also their understanding of astronomy. Ancient Greece, particularly Athens, was a crucible of philosophical thought. Great philosophers like Socrates, Plato, and Aristotle laid the groundwork for Western philosophy. They grappled with questions about ethics, politics, and the nature of reality.

The following centuries witnessed the rise of great empires such as Rome and China. During these times, knowledge became more structured, leading to the establishment of schools, libraries, and universities. These institutions became the repositories of human knowledge, facilitating its preservation and dissemination.

The Middle Ages marked a period of knowledge consolidation. Despite the common misconception of this era as a 'dark age', it was a time of significant intellectual activity. In the Islamic Golden Age (eighth to fourteenth centuries), Muslim scholars preserved and expanded upon the classical knowledge of Greece and Rome, making significant advances in areas like mathematics, medicine, and astronomy. Meanwhile, in the monasteries and universities of Europe, scholars were diligently copying and preserving ancient texts, preparing for the coming intellectual revival.

The Renaissance (fourteenth to seventeenth centuries) heralded a renewed thirst for knowledge. Inspired by the rediscovery of classical Greek and Roman knowledge, Renaissance scholars began to challenge conventional wisdom. The scientific revolution that accompanied this era led to a new, empirical approach to acquiring knowledge. Galileo's use of the telescope and Newton's laws of motion exemplified this shift.

The Industrial Revolution (eighteenth to nineteenth centuries) marked another critical period in the evolution of knowledge. As society became increasingly mechanised, the need for new knowledge and skills drove educational reform. The concept of compulsory education emerged, leading to a more literate population.

The twentieth century ushered in the age of digital information. The invention of computers and the Internet revolutionised the way knowledge is created, stored, and shared. Today, knowledge is often digital, accessible, and instant. With the click of a button, we can access an almost infinite amount of information.

The twenty-first century, up until the present day, has seen exponential growth in knowledge. Driven by technology and global connectivity, knowledge has become more specialised and diverse. It is no longer confined to textbooks or institutions but thrives in online communities, digital databases, and artificial intelligence. The era of big data and AI heralds an exciting new chapter in the story of knowledge, blurring the line between human and machine intelligence.

The Evolution of Knowledge in the Indian Subcontinent

The evolution of knowledge in the Indian subcontinent played a crucial role in shaping the world as we know it today, and it deserves separate attention.

The story of knowledge in the Indian subcontinent can be traced back to the ancient Harappan civilisation (around 2500 BCE). This Bronze Age culture exhibited advanced city planning and a system of standardised weights and measures, which speaks volumes about their mathematical knowledge. Unfortunately, their script remains undeciphered, leaving a gap in our understanding of their intellectual achievements.

In the centuries following the decline of the Harappan civilisation, the region saw a flowering of knowledge and culture with the advent of the Vedic period (1500-500 BCE). The four Vedas composed during this period form the bedrock of Hindu philosophy, ethics, rituals, and society. The intellectual pursuits were not merely spiritual but also extended to practical fields such as medicine (Ayurveda), mathematics, astronomy, and governance.

Around the sixth century BCE, a period known as the 'Axial Age', India saw the rise of significant philosophical schools of thought, including Buddhism and Jainism. This was also the period of the Upanishads, philosophical texts that contemplated the nature of reality and the self. They laid the foundation for the rich philosophical traditions that would continue to evolve in India for millennia.

India's golden age of knowledge, often referred to as the Gupta Period (fourth to sixth century AD), saw extraordinary developments in various fields. The concept of zero and the decimal system, the cornerstones of the modern number system, were invented during this time. Renowned scholars like Aryabhata and Varahamihira made significant contributions to astronomy and mathematics, and their works influenced thinkers around the world for centuries to come.

In literature and the arts, the Gupta period witnessed the composition of the famed epic poems, the *Mahabharata* and the *Ramayana*, which continue to influence South Asian literature and culture to this day. Furthermore, Kalidasa, the celebrated playwright and poet, composed his timeless works during this period.

The medieval period saw the synthesis of Hindu and Muslim traditions, with Persian, Arab, and Turkic influences enriching the Indian knowledge tradition. This era saw the development of regional languages and literature, advancements in architecture and music, and the evolution of complex administrative systems.

Fast forward to the Colonial Era (seventeenth to twentieth century), the Indian subcontinent experienced a confluence of Western and Indian knowledge systems. The introduction of the printing press and Western-style education, along with the translation of Indian texts into European languages, led to a new wave of intellectual exchange.

After independence, India has made significant strides in various fields, including space technology, information technology, biotechnology, and nuclear science. The growth of the Indian IT industry and the successful Mars and Moon missions by the Indian Space Research Organisation (ISRO) are testaments to India's contemporary intellectual prowess.

In essence, the evolution of knowledge in the Indian subcontinent has been dynamic and multifaceted, and it is a testament to India's intellectual heritage that many of its philosophical concepts, mathematical innovations, and artistic expressions continue to resonate with and influence global thought and culture today.

Knowledge and Technology: A Dynamic Symbiosis

From the invention of the wheel to the development of quantum computers, technology has been an integral part of human history. While technology has traditionally been viewed as a product of knowledge, it is increasingly clear that the relationship between technology and knowledge is symbiotic and profoundly interconnected.

At its core, technology can be defined as the practical application of knowledge for specific purposes. Every tool, device, or system we develop is based on our understanding of the world and our knowledge. For example, the creation of the steam engine during the Industrial Revolution stemmed from our knowledge of physics, specifically thermodynamics. Thus, technology is undoubtedly a product of knowledge.

However, the connection between technology and knowledge extends beyond this linear cause-and-effect relationship. As much as technology is a product of knowledge, it is also a tool for its acquisition, organisation, and dissemination. In other words, technology shapes knowledge.

Take the invention of writing systems around 3000 BCE, one of the most transformative technological innovations in human history. It enabled humans to record and pass on information, opening up vast new possibilities for the preservation and accumulation of knowledge. The written word was a technological tool that fundamentally transformed human knowledge and continues to shape it today.

The printing press, developed by Johannes Gutenberg in the 15th century, provides another example of how technology can revolutionise knowledge. This invention facilitated the mass production of books, breaking down barriers to knowledge access and paving the way for the intellectual awakening of the Renaissance and the Scientific Revolution.

In the modern era, the symbiosis between technology and knowledge is even more apparent. The advent of computers and the Internet has reshaped the landscape of knowledge in unprecedented ways. Today, a world of knowledge is just a click away, leading to the democratisation of information. Digital technology has accelerated the rate of knowledge production, facilitated global collaboration, and transformed the way we learn, think, and solve problems.

Artificial intelligence, the cutting-edge technology of the twenty-first century, exemplifies how technology has become an active participant in the creation of knowledge. These technologies don't just apply existing knowledge; they help generate new knowledge by identifying patterns, making predictions, and learning from data. The development of AI has even led us to reconsider the nature of knowledge and intelligence, traditionally seen as uniquely human domains.

Simultaneously, as technology becomes more integrated into our lives and societies, it shapes what we value as knowledge. In a digital society, skills such as coding, data analysis, and digital literacy are increasingly important. Furthermore, the omnipresence of technology raises new questions that require exploration and understanding, leading to the creation of new fields of study like cyber security, digital ethics, and AI governance.

As we continue to develop and adopt new technologies, we can expect this fascinating interplay to persist, leading to new forms of knowledge and novel technologies in a continuous cycle of innovation. The challenge and opportunity for us as a society lies in harnessing this symbiosis responsibly, equitably, and creatively for the betterment of humanity.

Knowledge and Education: Another Dynamic

The intersection of knowledge and education has long been a critical component of human progress. As we gaze into the future, emerging technologies and societal shifts promise to fundamentally transform how we disseminate and acquire knowledge.

The first two impacts will be that learning will become more personalised and a lifelong affair. As we accumulate more data about individual learning patterns and preferences, AI will allow for customised learning experiences that adapt, in real time, to a student's strengths and weaknesses. This will make learning more efficient, effective, and engaging. The rapid pace of technological and societal change will necessitate lifelong learning. As new technologies emerge and job markets evolve, the ability to continuously acquire new knowledge and skills will become essential. This will lead to an increased emphasis on self-directed learning and continuous professional development.

We have had occasion to talk at length about VR, AR, and the Metaverse, which will transform the way we learn by providing immersive, experiential learning opportunities. Imagine a history lesson where students can virtually walk through ancient Rome or a biology lesson where they can explore the human body from the inside. The Internet will continue to break down geographical barriers to education, creating global learning communities. Online platforms will facilitate collaboration and knowledge sharing across borders, fostering diversity and inclusivity in education.

Also, as the challenges facing our world become more complex, there will be a shift towards more multidisciplinary education. Students will be encouraged to make connections between different fields of knowledge, fostering creativity, critical thinking, and problem-solving skills.

With the rapid evolution of knowledge, the ability to learn how to learn—known as meta-learning—will become increasingly important. Education will not only focus on the acquisition of specific knowledge but also on developing the skills and strategies needed for effective learning. As AI systems become more sophisticated, they will not only serve as tools for learning but also as sources of new knowledge. Machine learning algorithms can identify patterns and generate insights that are beyond human capacity, contributing to the creation of new knowledge.

Most importantly, as we grapple with the social, political, and ethical implications of technological advancements, ethics will become an integral part of education. Students will need to learn to navigate complex issues like data privacy, AI ethics, and digital citizenship. It will be crucial to address issues of equity and access, ensuring that the benefits of these advancements are available to all. Embracing this future calls for a collective effort from educators, policymakers, technologists, and learners themselves.

Ubiquitous Access to Knowledge: Beyond Google

From encyclopaedias to libraries, we moved to Google and Wikipedia, transforming how we search, acquire, and interact with knowledge. However, the future holds even more extraordinary possibilities, some of which we have begun adopting today.

Imagine AI-powered personal assistants that understand your context, learning style, and preferences. These systems will not only fetch the information you need but present it in the most digestible and meaningful manner for you. We have spoken about ChatGPT, Bard, and the like.

IoT, too, will play a significant role in creating seamless access as connected devices around us will interact with each other, with AI at the helm, providing instant access to information and learning resources. Imagine your smart home understanding your child's school syllabus and helping them learn through interactive exercises and quizzes.

Perhaps the most extraordinary possibility is the use of brain-computer interfaces (BCIs) that could create a direct communication pathway between the brain and external devices. While the technology is still in its infancy, in the future, it might allow us to access and interact with digital knowledge through thought alone, making knowledge acquisition nearly as intuitive as thinking.

Blockchain technology could contribute to the development of decentralised knowledge networks, where information is shared peer-to-peer without the need for a centralised authority. This could make access to knowledge more democratic and resilient to censorship, while the development of quantum communication could revolutionise information transfer, making it more secure and efficient. Quantum internet might make it possible to connect quantum computers, leading to an exponential increase in computing power and the way we access and process information.

The Downside

While with such advancements would come the associated traditional challenges like data privacy, security, and the digital divide, there are other concerns as well, and the jury is out on how we would fare on these.

One is that children, compared to adults, are more susceptible and vulnerable in their formative years. The sheer amount of information and knowledge that their minds must deal with even today is mind-boggling. What will it be like in the future? So, information overload remains a key concern.

Similarly, unlike AI, humans have limitations when it comes to multitasking. We suffer from loss of concentration, reduced productivity, health issues, etc. when asked to cope with rapid changes around us in our work environment as well as what now constitutes 'leisure and entertainment'. Just looking at a screen for long hours in a day is enough to fatigue us. Will we survive the pace that AI will set for us?

Another big issue is isolation. Diminishing human contact in the exchange of knowledge and learning and an over-dependence on technology and AI also cause a growing sense of isolation in humans, leading to psychological problems.

All these factors can contribute to a loss of critical thinking skills and creativity rather than their enhancement.

■■■

In the future, we'll know everything there is to know. The downside? We'll forget where we left our smart glasses that tell us everything we know.

CHAPTER – 5

Humans and Food

The Malthusian Problem

In his 1798 book, *An Essay on the Principle of Population,* Thomas Malthus proposed that human population growth will always tend to outrun the food supply, inevitably resulting in famine and other catastrophes until the population is reduced below a sustainable level.

Malthus argued that population growth follows an exponential curve while the food supply grows at a linear rate. It meant that the population would eventually outstrip the food supply, leading to widespread starvation and death. He believed that there were only two ways to prevent this from happening—that the population be kept in check by natural disasters, wars, and diseases or that people choose to have fewer children, either by practising abstinence or by using birth control.

Malthus's theory was controversial at the time, and it is still debated today. Some people believe that his predictions were correct and that we are now facing a Malthusian crisis. Others believe that his theory is outdated and that technological advances have made it possible to increase the food supply faster than population growth.

The critics of Malthus' theory point out that the assumption that food supply only grows linearly is not always the case, as shown in recent decades due to technological advances in agriculture. Malthus did not anticipate the development of new technologies, such as fertilisers, pesticides, and irrigation, which have made it possible to produce more food on less land.

He also did not take into account the role of social factors in population growth. For example, the decline in birth rates in many developed countries is not due to natural disasters or wars but rather to changes in social norms and values.

The world's population is currently about eight billion people and is expected to reach 9.7 billion by 2050. Human population growth is expected to level off and eventually decline. However, the exact level at which the population will stabilise is uncertain. Some experts believe that the population will stabilise at around ten billion people, while others believe that it could stabilise at a much higher level, such as 15 billion or even 20 billion. So, are we likely to run out of food at some point in spite of all the advances in food tech to date?

The Impact of Climate Change

The real problem in predicting food supplies into the distant future is an unpredictable genie of our own creation—climate change.

Agriculture, more than anything else, has always depended on local weather patterns that do not deviate much in the long term. Soil conditions, availability of sunlight and water, and the timing of the seasons have long dictated crop choices and yields across the world. Uncertainty was always the bane of producing food, with boom-and-bust cycles, and economically, it remains a heavily subsidised sector in most parts of the world. In the global south, farmers can barely eke out an existence.

Bring climate change into the mix, the effects of which are clearly visible as we write, and you have taken uncertainty in food production to a whole new level. To feed the eight billion people, there are 100 billion plus farm animals causing dramatic climate change. While technology has so far successfully countered Malthusian naysayers, whether it can continue to do so in the face of an angry and capricious Nature will remain an open question for now.

However, that does not mean that advances in the future will not be pathbreaking and capable of overcoming at least some of the challenges that may arise.

IoT, Big Data, and AI in Food Tech

Technology has deeply impacted food in all domains—from planting and harvesting to packaging, distributing, and consuming food.

Robots and drones in agriculture are commonplace, at least in the developed world. Robots help plant various fruit and vegetable saplings that grow in nurseries and pick ripe fruits and vegetables that are delicate, thus avoiding any damage. They have also made inroads into sorting and grading food products, thus ensuring that only the best quality reaches the customers.

Analytics aids both in crop production and livestock maintenance. Inventory holding has become less wasteful, and costs have, thus, also reduced. Various tools help analyse consumer demands and know about emerging trends as well. This helps companies manufacture and stock the requisite resources accordingly. IoT sensors are being used to monitor crops and livestock in real time. This data can be used to optimise irrigation, fertilisation, and other agricultural practices. For example, IoT sensors can be used to track the moisture levels in soil, which can help farmers determine when to irrigate their crops. Technology also helps to identify food anomalies. Not only can tracking devices help in finding out the source of a food item in case it turns out to be rotten, but it is also able to find out why a particular crop has gone awry. Sensors are being used to preserve food for longer by helping with temperature control while storing and transporting food products. Drones are being used to deliver food faster in last-mile connectivity to the consumer.

Big data is being used to analyse data from IoT sensors, as well as other sources, such as social media and weather data. This data can be used to identify trends and patterns that can help farmers improve their yields and reduce their costs.

AI is being used to automate tasks in food production, such as crop monitoring and pest control. For example, AI can be used to analyse images from IoT sensors to detect pests and diseases. This helps farmers in taking action to prevent crop losses.

Vertical Farming

Vertical farming is a type of agriculture that grows crops in vertically stacked layers. This method of farming uses less land and water than traditional agriculture, and it can be located in urban areas, making it more accessible to consumers. It also helps to reduce the environmental impact of food production and makes it possible to produce fresh, local food that is not transported over long distances. The concept of vertical farming was first proposed in 1999 by Dickson Despommier, professor of Public and Environmental Health at Columbia University.

Vertical farming can be of different types—hydroponics involves growing plants in liquid solutions containing macronutrients like nitrogen and trace elements like iron, aquaponics involves farming with the production of aquatic organisms in a closed-loop system, and aeroponics was motivated by the initiative of NASA, which aimed to grow plants in space.

Despite challenges, there is a lot of potential for vertical farming to become a major part of the food production system in the future. As the technology continues to develop and the cost of vertical farms comes down, this method of farming is likely to become more widespread.

GM Crops of the Future

Genetically modified (GM) crops are a type of crop that has had its DNA altered in a way that does not occur naturally. This can be done to improve the crop's resistance to pests, diseases, or herbicides or to improve its nutritional content. They can also be made resistant to climate change, less water-dependent, give a higher yield, or tailored to specific needs like being gluten-free.

GM crops have been controversial since their introduction in the 1990s. Some people believe that they are safe and can be a valuable tool for improving food production, while others are concerned about the potential risks of GM crops, such as the possibility that they could harm human health or the environment.

Public acceptance of GM crops is still a challenge in some parts of the world. This is due to the concerns that some people have about the safety of GM crops. There are also regulatory hurdles that need to be overcome before GM crops can be widely adopted, which vary from country to country.

Despite these challenges, there is a lot of potential for GM crops to improve food production and human health in the future. As the technology continues to develop and the public becomes more accepting of GM crops, it is likely that we will see even more widespread adoption of GM crops in the years to come.

New Food

Lab-grown food, also known as cultured or cell-cultivated food, is shaping up to be a game-changer in the way we perceive and produce food.

It's a novel concept where food is produced through a process that involves the extraction of cells from a living organism. These cells are then cultured in a controlled environment, providing them with the nutrients and conditions they need to grow and multiply. The process eliminates the need for traditional farming, significantly reducing the need for land, water, and other resources.

Cultured meat is one of the most prominent examples of lab-grown food. Scientists take a small sample of animal cells (usually muscle cells) and place them in a nutrient-rich culture medium. The cells grow and multiply, eventually forming muscle tissue that can be harvested and processed like traditional meat.

The technology is not limited to meat; it also includes dairy, eggs, and other animal-based products. Using fermentation techniques and specially designed yeast, scientists can produce proteins identical to those found in milk or eggs. This approach can lead to completely vegan products that mimic the taste and texture of their animal-derived counterparts.

Apart from the environmental benefits, cultured meat can provide an ethical solution to animal farming. Since the meat is produced from cells, no animals need to be slaughtered, addressing concerns related to animal welfare. Since lab-grown foods can be produced in sterile environments, it reduces the risk of contamination and producers have control over the nutritional content, potentially creating products tailored to specific dietary needs.

Despite its potential, there are challenges to overcome—high costs, regulatory hurdles, and consumer acceptance—exactly as in the case of GM crops.

As research progresses and technology becomes more refined, lab-grown food may become a common sight on our plates, providing an ethical and sustainable alternative to traditional agriculture. Only time will reveal the full impact of this innovation, but the possibilities are as tantalising as they are promising.

3D Printed Food—Culinary Indulgence

Food connoisseurs are always on the lookout for something new. They are never satiated, and food companies know this. So, the ultimate culinary indulgence—food 'printed' right to your last specification using 3D printing technology.

The concept of 3D printing, also known as additive manufacturing, has been around for a while. It involves building objects layer by layer based on a digital model. When applied to food, this process uses various edible materials like 'ink', which are then deposited according to a preset design to create food in various shapes, textures, and flavours.

The 'inks' can range from plant-based substances, pureed foods, chocolate, dough, or a mix of ingredients loaded into the printer in a semi-liquid form. The 3D printer follows a digital blueprint, precisely placing the food material layer by layer to create the final product.

One of the most significant benefits of 3D food printing is the potential for personalised nutrition. With 3D food printers, meals can be customised according to individual nutritional needs, including adjusting carbohydrate, protein, or vitamin levels. This can be particularly beneficial for individuals with specific dietary restrictions or needs, such as athletes, the elderly, or those with chronic diseases.

While the journey from a prototype to a kitchen countertop device might be long and fraught with challenges, the potential benefits could make this culinary innovation worth the wait. As we continue to push the boundaries of what's possible, who knows what delicious possibilities await? As they say, the proof of the pudding is in the eating, or in this case, in the printing.

Nutraceuticals and the Magic Pill

The idea of a magic pill, one that contains all the nutrition our bodies require, is a notion that has captivated humanity's imagination for a long time. It's an enticing concept, especially in an era where time is of the essence, and convenience is highly sought after. Nutraceuticals have been touted as a potential precursor to this futuristic idea, but how close are we to this reality?

Coined from the words 'nutrition' and 'pharmaceutical', nutraceuticals are products derived from food sources purported to provide extra health benefits beyond basic nutritional value. They encompass a broad range of products, including fortified foods, dietary supplements, and herbal products, and are designed to improve health, delay ageing, prevent chronic diseases, or enhance life quality and longevity.

The concept of replacing meals with a single pill holds various attractions. It promises the convenience of eliminating time-consuming meal preparation, the certainty of receiving all necessary nutrients, and the potential for personalisation to individual dietary needs.

In theory, such a pill could contain proteins, fats, carbohydrates, fibre, and necessary vitamins and minerals, all tailored to a person's nutritional requirements. Advances in nutraceuticals, particularly in targeted delivery systems and bioavailability enhancement, may edge us closer to this reality.

While the concept is fascinating, several significant challenges make the idea of a complete diet in one pill more science fiction than fact at present.

Firstly, human nutritional needs are complex and not fully understood. There's more to food than just nutrients—there are elements like phytochemicals and antioxidants, and we're still learning about their roles. Plus, nutrients often work synergistically, meaning they can enhance each other's functions when consumed together in food.

Secondly, eating stimulates the digestive process and the release of enzymes, hormones, and other compounds. A single pill might not provide the same stimulation, potentially affecting gut health and overall digestion.

Thirdly, eating is not just about nutrition; it's a social and pleasurable activity. A pill cannot replicate the satisfaction derived from biting into a juicy apple or the joy of a shared meal.

And lastly, developing a pill that contains all necessary nutrients in the right proportions would be an enormous scientific challenge. It would also raise safety and regulatory issues, particularly concerning long-term use and effects.

However, the exploration of this idea has led to exciting advances in personalised nutrition, disease prevention, and health optimisation through nutraceuticals. As science continues to advance, we might not get that magical meal-replacing pill, but we will undoubtedly see innovations that push the boundaries of customised nutrition and wellness, making a future filled with health and vitality more attainable for all.

CHAPTER – 6

Humans and Medicine

Unending Evolution

Medical science is a field of unending evolution. From the ancient methods of healing to today's highly sophisticated techniques, the journey has been nothing short of miraculous.

Advancements in the last century have turned many fatal diseases into manageable conditions and transformed the healthcare landscape.

If we were to look at only three groundbreaking discoveries of the past, these would be the discovery of penicillin by Alexander Fleming in 1928, which marked the beginning of the antibiotic era, the invention of vaccines, specifically the smallpox vaccine, and imaging technologies.

Antibiotics revolutionised medicine, providing a powerful tool to combat bacterial infections that were often fatal before.

Edward Jenner's smallpox vaccine, followed by many more, has saved countless lives and eradicated or controlled many once-lethal diseases. The development of vaccines represents one of the most significant achievements in public health.

From the discovery of X-rays by Wilhelm Conrad Roentgen in 1895 to the advent of Magnetic Resonance Imaging (MRI) and Computed Tomography (CT) scans, imaging technologies have transformed diagnostics, enabling doctors to view the human body without invasive procedures.

Since then, it has been a ceaseless march of innovation, invention, and discovery overlapping with many other fields, and the completion of the Human Genome Project in 2003, which mapped the entire human genetic code, has opened doors to understanding diseases at a molecular level and laid the foundation for personalised medicine.

Groundbreaking Innovations—Present Day

Medicine continues to advance rapidly.

One is telemedicine, where the integration of digital technology with healthcare has enabled remote patient monitoring and virtual consultations, improving access to healthcare services. AR and VR have significant roles in improving telemedicine outreach. For example, AR and VR are being used to provide surgeons with a more immersive and interactive experience during remote surgery. They are being used to create realistic and interactive simulations for medical training as well as patient education or mental health therapies. As more and more people are using their smartphones and tablets to access healthcare services, this trend is only going to continue, and it will make telemedicine even more convenient for patients.

Closely related to advances in telemedicine is remote patient monitoring, which allows doctors to monitor the health of their patients remotely, which is made possible by breakthroughs and advancements in wearable devices. These devices can collect a variety of health data, such as heart rate, blood pressure, blood glucose levels, and sleep patterns, which can be transmitted to healthcare providers in real time, allowing them to monitor patients' health remotely.

Apart from wearables, a huge amount of data is generated by modern diagnostic machines, scanners, laboratory tests, clinical trials, etc. Collecting such huge amounts of data with the help of IoTs and processing it using AI and machine learning helps in disease diagnosis, drug discovery, and personalised treatment, among other applications.

In contrast to traditional treatments like chemotherapy, newer methods like immunotherapy are being used to harness the body's immune system to fight diseases, particularly cancer. It represents a significant shift in cancer treatment and has proven effective even in some cancers previously considered untreatable. Since it can be tailored to the individual patient's immune system, the patients are more likely to respond to treatment and have better outcomes.

Genetic Engineering techniques like clustered regulatory interspaced short palindromic repeats and CRISPR-associated protein 9 (CRISPR-Cas9) are allowing precise genetic editing, holding the potential for curing genetic disorders and enhancing preventive medicine. It is being used to create vaccines that are more effective and safer than traditional vaccines. This is because CRISPR-Cas9 can be used to target specific pathogens and to insert genes that code for antibodies into cells. A Nobel laureate, Jennifer Doudna, who is known primarily for CRISPR, describes it as 'a technology that literally enables the rewriting of the code of life'. The work for which Doudna shared the Nobel Prize was published more than a decade ago, in 2012, opening up what seemed like an almost limitless horizon for CRISPR-powered therapies and cures.

But surveying the recent landscape of scientific breakthroughs, she says that we're still at an extraordinary time of accelerating discoveries.

Similarly, the development of messenger Ribonucleic Acid (mRNA) vaccines during the Covid period captures the compression of drug-development and vaccine development timelines from decades to weeks and months, thus saving millions of lives across the world.

A World of Possibilities—Future Day

The future holds many promising areas where research from diverse fields is being brought to medicine and put to cross-functional use. We have already seen in the previous section the use of artificial intelligence (AI) in medicine, the use of wearable devices, and the gene editing technology CRISPR-Cas9, which are all examples of cross-domain applications of technology. Similarly, 3D printing is being used to create customised medical devices, such as prosthetics and implants. Some of the most promising developments are occurring in the domains below.

Regenerative Medicine

Regenerative medicine represents a radical shift in healthcare, an innovative approach that aims not only to treat disease but to regenerate tissues and organs, restoring or establishing normal function. It has been hailed as the next big leap in medical science, promising cures for previously untreatable conditions and injuries.

Regenerative medicine focuses on the body's regenerative capabilities to treat illnesses and injuries, using methods to repair, replace, or regenerate human cells, tissues, or organs.

The primary goal is to restore or establish normal function in conditions where the body's self-healing processes aren't sufficient.

This field encompasses several strategies, including cellular therapies (transplanting cells or stem cells to stimulate the body's repair system), tissue engineering (creating functional tissues in a laboratory to replace or restore damaged tissues in the body), medical devices, and artificial organs (that replace the function of a damaged organ or supplement a specific function to allow the organ to rest and recover).

Stem cells have the unique ability to develop into many different cell types and replicate rapidly. They serve as a sort of internal repair system in many tissues, dividing essentially without limit to replenish other cells. Stem cells can be harvested from various sources, including bone marrow, adipose tissue (fat cells), and blood, but the most potent stem cells come from embryos (embryonic stem cells). They have been utilised to treat a variety of diseases, such as heart disease, neurodegenerative diseases, and diabetes. They also offer significant potential in treating spinal cord injuries and restoring vision to the blind.

An excellent example is the creation of lab-grown skin for burn victims or patients with non-healing wounds. Future possibilities include engineered heart valves, blood vessels, liver tissue, and even entire organs.

Biomedical engineering integrates principles of engineering with the biological and medical sciences to develop devices and equipment that enhance patient care. This includes developing artificial organs or medical devices that can perform the functions of the organ they replace. Prosthetic limbs have become increasingly sophisticated, often employing advanced materials and electronics to mimic the function of natural limbs.

Nanomedicine

Another emerging frontier is the application of nanotechnology in medicine, including targeted drug delivery and nanorobots, which can revolutionise diagnostics, treatment delivery, and disease monitoring.

Nanomedicine integrates knowledge from various fields, such as physics, biology, engineering, and medicine. It utilises nanoscale materials, typically in the range of one to 100 nanometres (a nanometre is one billionth of a metre), to diagnose, treat, and prevent diseases. For perspective, a single human hair is approximately 80,000–100,000 nanometres wide. The small size, unique physical properties, and high surface-to-volume ratio of nanoparticles make them ideal for a variety of medical applications.

One of the most promising applications of nanomedicine is the targeted delivery of drugs. Nanoparticles can be designed to carry therapeutic agents and deliver them precisely to the disease site, minimising the exposure of healthy cells and reducing side effects. For example, in cancer treatment, nanoparticles can be engineered to carry chemotherapy drugs directly to tumour cells, sparing healthy cells. This approach can enhance the efficacy of treatment while significantly reducing side effects.

Nanotechnology can enhance the sensitivity and precision of diagnostic procedures. Nanoscale contrast agents can improve imaging from techniques like MRI, leading to earlier and more accurate detection of diseases. Additionally, nanoparticles can be engineered to bind to specific biomarkers on diseased cells. They can be visualised using advanced imaging techniques, allowing for more precise disease diagnosis and monitoring.

Nanotechnology has a significant role in tissue engineering and regenerative medicine as well. For instance, nanofibrous scaffolds can mimic the natural environment of cells, assisting in tissue regeneration and healing.

Precision Medicine

Riding on the wave of genomics and big data, precision medicine allows for treatments tailored to individual genetic and biochemical makeup, ensuring the most effective interventions.

Historically, medical treatments have been developed for the 'average' patient based on what works best for most people. However, the 'one size fits all' approach doesn't take into account the genetic, environmental, and lifestyle differences that make each person unique. Precision medicine challenges this approach by taking these factors into account, recognising that each patient is unique and may respond differently to treatments. The goal is to optimise therapeutic effectiveness for particular groups of patients, thereby improving outcomes and reducing side effects.

The completion of the Human Genome Project in 2003 was a milestone that laid the foundation for precision medicine. Genomic information can identify genetic mutations that increase the risk of diseases, predict how a patient will respond to a drug, and even suggest preventive measures for at-risk individuals. Genetic testing, through technologies like next-generation sequencing, can identify inherited disorders, confirm a diagnosis, predict future diseases, and guide doctors in choosing the best treatments for a particular patient.

Precision medicine also leverages health information from electronic health records (EHRs), wearables, and other digital health tools. When combined with genomic information, this data provides a comprehensive picture of a patient's health status. AI and machine learning play a significant role in analysing this 'big data', uncovering patterns and insights that can guide individualised treatment strategies.

Keeping the Dead Alive—Cryonics

For centuries, humankind has sought the secret of immortality, pursuing elixirs of life, fountains of youth, and countless other myths and legends. While eternal life remains beyond our grasp, a modern scientific technique known as cryonics seeks to offer the next best thing: the chance to extend life far beyond its natural limits.

Cryonics, a term coined in the early 1960s by Robert Ettinger in his book *The Prospect of Immortality*, involves preserving human bodies at ultra-cold temperatures with the hope that future medical and technological advancements could revive them and cure currently incurable diseases.

The idea behind cryonics is to 'pause' life in a state where it can be held indefinitely and 'resumed' at a time when technology has advanced significantly. In its current stage, cryonics is a form of medical 'time travel', a speculative life support that reaches out to a future era of unprecedented medical capabilities.

The process of reviving a cryopreserved body while ensuring that the brain, which holds personality and memory, is not damaged is still theoretical. No mammal has been cryopreserved and revived to date.

Two methods of brain preservation that have gained attention are Aldehyde-Stabilized Cryopreservation (ASC) and chemopreservation. ASC involves using a fixative (like glutaraldehyde) to stabilise the brain and then cryopreserve it at extremely low temperatures. This method was used to win the Brain Preservation Prize in 2016, demonstrating that an entire rabbit brain's connectome (the map of neural connections) could be preserved using this technique. While it's not currently reversible (the process is damaging and not designed for biological revival), it has been proposed for use in a theoretical process called 'mind uploading', where the preserved information could be used to recreate the individual's mind in a computer simulation or different biological or synthetic substrate.

Chemopreservation involves using chemical fixatives to preserve the brain at room temperature. The idea here is that, rather than preserving the brain for future biological revival, the goal is to preserve the brain's information content for potential future scanning and mind uploading. This method is more speculative and isn't as widely pursued as cryonics or ASC.

It's important to note that these preservation methods are still largely theoretical and controversial. They depend on future technologies for revival or mind uploading, neither of which currently exist, and the process of preservation is considered irreversible with current technology.

Legal systems around the world differ in their recognition and regulation of cryonics. Additionally, ethical concerns include issues around informed consent, the definition of death, and religious and philosophical beliefs about life and death.

Cloning

Creating an exact genetic replica of another cell, tissue, or organism is called cloning, and the replica that is created is the clone. Remember Dolly, the Scottish sheep?

Cloning can be of different types. Gene cloning is where replicas of genes or segments of DNA are made; in reproductive cloning, whole copies are made; and therapeutic cloning creates embryonic stem cells to be used to treat injuries and diseases. The most common of these are therapeutic and reproductive cloning.

As of today, therapeutic cloning is not in medical practice anywhere in the world.

Reproductive cloning involves the creation of an entire cloned human. The possibility of this, and much speculation about it, has been around for the greater part of the twentieth century and continues. The term was first used in 1969 by JBS Haldane. However, animal/organism cloning was done much before by Hans Driesch in the late 1800s by splitting a sea urchin embryo and then with a salamander in 1902. In 1952, Robert Briggs and Thomas Joseph King cloned a frog embryo. But the major breakthrough came with the making of Dolly in 1996. After this, the possibility of making human clones increased. A hybrid human clone was actually created in November 1998 by Advanced Cell Technology, where a nucleus was taken from a man's leg cell and inserted into a cow's egg from which the nucleus had been removed. This hybrid cell was cultured and developed into an embryo. The embryo was, however, destroyed after 12 days.

There is a world of debate over cloning, like all other cutting-edge advances, including its legal and ethical implications and its impact on God's role as the Creator.

Future Anxieties—Peeping Through the Black Mirror

Teleportation Paranoia: The fear that you'll materialise somewhere without your pants.

Space Tourist Anxiety: The worry that you left the oven on back on Earth.

Robot Pet Attachment: Concern that your robot dog likes the neighbour's robot cat more than you.

AI Overthink: Anxiety that your smart home might break up with you because you didn't say 'please' when asking to adjust the thermostat.

Mind-Reading Mishaps: Worrying that people will discover your secret love for '80s disco music when brain-to-brain communication becomes a thing.

Hologram Etiquette Stress: The anxiety of not knowing how to politely 'end call' on a too-realistic holographic meeting.

Time Traveller's Guilt: Concern that you should be doing something productive, like stopping a historical event, instead of binge-watching shows from the future.

Biohacking Blunders: Worrying that upgrading your body firmware might introduce new bugs, like a sudden fondness for interpretive dance.

Alien Social Pressure: The fear that extraterrestrials will unfollow you on Intergalactic Instagram because your Earth selfies are 'basic'.

CHAPTER – 7

Humans and Society

Living with Each Other

Humans have lived with humans for millennia, mostly in bands and tribes. As agriculture-based societies, they have lived together for a few thousand years; as industry-based societies, they have lived with (increasingly sophisticated) machines for a few centuries; as an information and data-based society, for a few decades; and now, humans have just begun to live with AI.

Most of our social structures are based on the first fact. Questions of race, caste, creed, nationality, and religion that came into existence long ago remain fundamental (and unanswered) today. While we have whole academic disciplines, centuries of research (and a lot of terminology) explaining these 'constructs', and a lot of 'affirmative' action to solve these by governments and 'civil societies' across the world, it does not take long for a teenager maturing into adulthood today to figure out that these 'constructs' continue to haunt and determine their future. African Americans in the USA and Dalits in India continue to face the same challenges as their parents and grandparents—something quite discouraging for 'progressive' societies. A lot of 'progress' in 'technology' has been invested in perpetuating the established 'constructs' and fighting old battles.

In this sense, it is clear that technology per se has no 'agency'; it is a means to an end, and unfortunately, humans as a society have once too often failed to act for the 'greater good'. More machines and more information have more often than not translated into more powerful means to weaponise society to harm the 'other'. Far from a 'liberator', technology has equally proved to be an 'enslaver'.

1984 and the Politics of the Global Village

Orwel's dystopian vision where Big Brother was constantly watching was in the realm of fiction, which is no longer the case because today, technology makes it all too possible with Big Data, IoT, and AI. From 'closed' societies like China to 'open' societies like the US, surveillance is ubiquitous. Private companies hoard digital footprints for a reason; they can be exploited later for profit or an agenda.

In the global village that the world has been for a while now, thanks to multiple social media platforms where the chatter never ceases, conversations easily descend to the lowest common denominator, victims of the same biases and prejudices. Our politics of the future appears to be a continuation of the present.

Throughout history, propaganda—biased or misleading information used to promote a political cause or point of view—has been a tool leveraged by groups aiming to shape public opinion. However, tech and AI have ushered in a new era of propaganda that is more pervasive and harder to detect than ever before.

Social media platforms have become fertile ground for spreading propaganda due to their extensive reach and the sheer volume of content they host. Through cleverly crafted posts, images, and videos, manipulative narratives can be disseminated to millions of people within a matter of minutes. A key factor in this process is the creation of echo chambers—online spaces where individuals are exposed to opinions that align with their own, reinforcing their existing beliefs and making them more susceptible to certain types of propaganda. Social media algorithms, which suggest content based on user behaviour, often inadvertently create and strengthen these echo chambers.

AI has added another layer of complexity to the problem. Deepfakes, or synthetically created realistic videos, audio clips, or images, are becoming increasingly easy to create thanks to advances in AI. These can make it appear as if politicians or public figures are saying or doing things they never did, leading to public misinformation and potential political manipulation.

If we had a measure of global societal and political cohesion, say on a scale of ten, and if we could plot where we were in the past and where we stand today, the needle would not have budged much, and it would seem, extrapolating to the future, even if we are an inter-planetary species tomorrow, we may not be any more 'politically cohesive' than we are today, and the needle may not move much further either.

On the impact technology is having on the individual psyche, the changes are visible and massive, too. Put together, these changes will result in other social changes, such as the idea of marriage and companionship, problems like stress and loneliness, ageing, etc., which we examine in more detail.

Caught in the Web

The quickest to adopt and adapt to new tech are the young—children and young adults. The problem is that they are also the most susceptible to influence and moulding and to unfiltered information, manipulative commerce, and illegal exploitation—a trend most visible on social platforms.

Given the fact that social platforms are not just tools but an essential part of their social ecosystem, there is no wishing them away. The three areas in which social media contributes positively—learning and education, socialisation, and self-expression and identity formation—are real, tangible benefits.

Platforms like YouTube offer countless educational videos on a range of topics, helping youngsters supplement their formal education and pursue personal interests. For young adults, platforms like LinkedIn provide networking opportunities and career development resources. Young people connect with friends and make new ones on social platforms regardless of geographical limitations. It can also provide a sense of belonging, particularly for those who feel marginalised or misunderstood in their physical communities. Social media offers a platform for self-expression, helping children and young adults explore and shape their identities. It can serve as a digital canvas where they can share their thoughts, ideas, creativity, and experiences.

But the pitfalls are big, real, and deserve our attention, too. Harmful behaviours such as cyberbullying under the cloak of anonymity that the Internet sometimes provides lead individuals to make hurtful comments or share damaging content about others, leading to severe emotional harm and mental health problems.

Excessive screen time and social media use lead to addiction, resulting in sleep disturbances, poor academic performance, and reduced physical activity. Additionally, it leads to harmful effects on mental health, contributing to anxiety, depression, and low self-esteem.

Platforms like Instagram, known for curated content and highly edited images, propagate unrealistic beauty and lifestyle standards, leading young people to make damaging comparisons and contributing to body image issues and poor self-esteem.

Also, there are privacy and security risks. Younger users may not fully understand the implications of sharing personal information online. This naivety can expose them to exploiters, including data theft, online stalking, and inappropriate contact with strangers. Nations are trying to bring some 'sanity into this madness' through regulation and laws. An example of this is the European Union's General Data Protection Regulation, which focuses on the obligations of data controllers and processors to provide individuals with access to their data and information about how it is being used.

Metaverse Menace

If you thought social media was bad, the Metaverse could take this 'devolution' to another level. We might become prisoners of meta platforms, controlled by the tech lords, sometimes without knowledge, and mostly without consent. We could be programmed to crave the experience and paraphernalia—gaming, social interactions, shopping, and virtual developments.

The Metaverse seeks to combine simulated sights, sounds, and feelings into what we know as the real world, thus, changing and even distorting how we interpret our daily life. It aims to merge the physical and the virtual world so compellingly that boundaries will vanish—we will believe that certain people, places, and objects exist when they don't. The Metaverse might provide an escape from reality, but it might end up being manipulation at its best.

The most significant conundrum that will lie in front of us would be how to exit the Metaverse. In the 2D world, we could turn off our phones and computers and move away from what we don't like. But with the sheer psychological power of the Metaverse, you would always remain a part of it.

The New Dynamics of Personal Relationships

Social media has fundamentally changed the way we form and maintain relationships. On the one hand, it has made it easier to stay in touch with loved ones, share life updates, and foster connections with people around the globe. As for romantic relationships, it allows couples to stay connected throughout the day, sharing moments, messages, and feelings in real time.

However, the hyper-connectivity enabled by social media can also create challenges. For marriages and relationships, over-reliance on virtual communication can sometimes supplant face-to-face interactions, leading to feelings of disconnection and emotional distance. The phenomenon of 'phubbing' (snubbing someone in favour of your phone) can lead to reduced relationship satisfaction and increased conflict. People in the same room are not together; they are always 'elsewhere'. Unfortunately, tech distractions, like mobile calls and notifications, take priority over humans around us. When we are travelling for work, we call family and friends because we miss them, but when we are with family, we take work calls. We are living less and less in the moment.

Moreover, the curated nature of social media content can lead to unrealistic comparisons, leading individuals to question the quality of their relationships or lifestyles. This phenomenon, known as 'comparison despair', sows discontent and exacerbates feelings of inadequacy or loneliness.

Artificial intelligence is also shaping our interpersonal dynamics. AI can facilitate connections with dating apps using complex algorithms to match potential partners based on shared interests, values, and other compatibility factors. This data-driven approach can streamline the search for a partner, making dating more efficient.

However, there are also potential downsides. The algorithmic nature of these platforms can lead to a transactional approach to dating, where individuals are swiped away based on superficial criteria. This can lead to a lack of depth and authenticity in connections formed, impacting the quality of relationships.

AI is also being used to create virtual companions, from chatbots to more sophisticated AI-powered robots. While these companions can provide comfort and reduce feelings of loneliness, they also raise ethical and psychological questions. Over-reliance on AI companions could potentially lead to social isolation as human relationships are replaced by artificial ones.

While our digital tools are becoming more sophisticated, rates of loneliness and social isolation are on the rise. This paradox underscores the fact that the quantity of connections does not equate to quality. Real, meaningful relationships require more than just digital interaction—they require empathy, understanding, and shared experiences.

To navigate this new digital landscape, we need to foster a healthy balance between our online and offline lives. Digital literacy education would have to include not just the technical aspects of using these tools but also their psychological and societal implications. Importantly, we must also ensure that our digital technologies are designed and used in ways that promote genuine connection and well-being.

A Grey Future

As societies become more digitally integrated, profound changes are occurring in our demographic structures as well. Women are marrying later and having fewer children, leading to an ageing population in many developed societies.

Many factors have contributed to this trend, including increased access to education and career opportunities for women, the rise of more effective contraceptives, and shifting societal values around marriage and parenthood. Social media and dating apps have transformed how relationships form, offering greater choice and flexibility. Consequently, people may spend more time dating and less time in committed marital relationships. Career-focused social networks like LinkedIn further contribute to this trend, allowing individuals to invest more in their professional development.

Moreover, the ubiquity of information online has led to increased awareness about the economic costs and lifestyle changes associated with raising children. With financial stability being a significant concern for many, particularly in uncertain economic times, couples may choose to delay having children or decide against it altogether.

As a result of these shifts, we see the rise of ageing societies. With fewer young people and an increased number of older adults, there are significant implications for economic growth, healthcare systems, and social structures.

The challenge for governments and societies is to develop sustainable models that can support an ageing population. This could involve rethinking retirement, promoting lifelong learning to keep older adults engaged and productive, and investing in healthcare innovations to manage age-related health issues.

AI and technology will play a crucial role in shaping the future of these demographic shifts. AI has the potential to mitigate some challenges associated with ageing societies. For instance, AI-powered health solutions could enable early detection and treatment of diseases, improving the quality of life for older individuals and reducing healthcare costs. Robotic caregivers could assist the elderly, providing companionship and support with daily tasks.

In terms of fertility, advancements in reproductive technologies, like in-vitro fertilisation (IVF) and egg freezing, are offering women more choices regarding when they have children. As these technologies continue to improve, it's likely that we'll see an increased decoupling of fertility decisions from biological age.

Despite these advancements, societies will still need to address the fundamental issues associated with ageing populations and decreased fertility rates. This could involve creating family-friendly policies that make it easier to balance work and parenthood, redefining societal expectations around age and productivity, and fostering intergenerational relationships and cooperation.

We are at a crossroads of demographic shifts influenced by societal changes and technological advancements. The challenge lies in harnessing these changes in ways that promote societal well-being and sustainability, ensuring that our digital future is not only more connected but also more caring and inclusive.

Navigating the Shadows—Cybercrime and the Darknet

One of the most troubling aspects of the digital age is the surge in cybercrime. These digital misdemeanours range from identity theft and phishing to more sophisticated crimes like ransomware attacks and corporate data breaches. As our lives become more digitally intertwined, the opportunities for these crimes increase exponentially. The most worrisome part is that many of these activities take place in the less-visible corners of the Internet, known as the Darknet.

The Darknet is a part of the Internet that isn't indexed by search engines and where identities are often hidden, providing a haven for illegal activities. It facilitates a vast and disturbing underworld marketplace where everything from drugs and weapons to stolen data and hacking tools is traded.

The repercussions of our evolving digital landscape aren't limited to the covert world of cybercriminals. They're also being felt in our everyday lives, particularly in how we interact with each other and perceive ourselves. One particular area of concern, especially for families with young children, is the ease with which pornographic material can be accessed and downloaded. The Internet provides users with the ability to download explicit images, engage in sexually suggestive conversations on online forums, play erotic games, and even purchase tickets for adult-oriented events. Furthermore, sexual predators have been known to use the Internet to identify and target potential victims.

Moreover, there's a growing concern about the role of AI in spreading misinformation and deepfakes (convincing fake videos created using AI), which have enormous potential for creating social discord and impacting political processes, as we have already seen.

Cyberterrorism is another danger—it refers to the exploitation of the Internet to execute violent actions that cause or intend to cause significant bodily harm or loss of life to further political or ideological objectives via threats or intimidation. Various techniques, including the use of computer viruses, worms, phishing, malicious software, hardware methods, and programming scripts, can all constitute forms of Internet terrorism, particularly in the context of purposeful and large-scale disruption of computer networks, including personal computers connected to the Internet.

Numerous instances of cyberterrorism, both major and minor, have occurred. In 1997, the National Security Agency (NSA) conducted an experiment in which 35 hackers were able to gain access to critical Pentagon computer systems, allowing them to modify accounts, reformat data, and even shut down entire systems. Al-Qaeda has also utilised the Internet as a means of communication and recruitment. In April 2007, Estonia, a technologically advanced Baltic country, became a battleground for cyberterrorism after a dispute over the relocation of a WWII Soviet statue in the capital city of Tallinn. More recently, in January 2023, Latvian officials claimed that hackers linked to Russia launched a cyber-espionage phishing campaign against the country's Ministry of Defence, although this operation was ultimately unsuccessful. In the same month, hackers disrupted access to more than 1,500 Nepalese government websites by overwhelming the main government server with traffic.

As the Internet continues to grow and computer systems become increasingly complex and interdependent, the potential for sabotage or terrorism via the Internet may become a more significant threat. Some experts even consider it to be one of the top ten events that could lead to the end of humanity.

CHAPTER – 8

Humans at War

Humans are a Warring Species

We began our last chapter on societal changes with the observation that our most deep-rooted biases, prejudices, customs, beliefs, and practices that prevent us from being 'one humanity' on the one planet that supports life have not changed or diminished in the face of astounding progress in every other sphere—science, technology, academic research, and our ability to introspect and understand our problems.

We know it is possible for the Earth to sustain us all in a way that all can thrive. We have the data but not the will to let the 'other' thrive. As a result, every tool, technology, resource and idea is 'weaponised' to serve our overt and covert national and sectarian goals.

We are fundamentally a warring species, and from being occasionally at war to mostly at war, we have finally earned the dubious distinction of being at war 24x7x365, thanks to our technological prowess and AI. From having soldiers fight soldiers to soldiers fighting civilians to machines fighting soldiers and civilians to now machines fighting machines, we are proving that killing is the thing that comes most naturally to us.

The ethical debates today are not about whether it is okay to kill another human but whether it is essential to have a human behind the kill button of a drone or whether it would be all right for the machine or AI to make a kill decision autonomously.

If science fiction and current debates and developments are any guide, the wars of the future will be all-encompassing, colossal affairs, and they will take us closer than ever, on the doomsday clock, to wiping ourselves off the face of this planet. That is the course we have set for ourselves.

Lethal Autonomous Weapon Systems (LAWS)

One of the most significant impacts of AI on warfare is the advent of autonomous weapons systems. These systems, which include drones and unmanned vehicles, can operate with minimal human intervention. They can perform tasks such as reconnaissance, target identification, and even launching attacks. The advantages are clear: fewer soldiers are put in harm's way, and operations can be more precise and efficient.

The notion of autonomous weapons is not new. Simple forms of fire-and-forget weapons, such as heat-seeking missiles, which could lock onto a target and pursue it without further guidance, have been in use for decades. However, these systems had limited autonomy and were used in relatively controlled environments.

The advancement in computer science, robotics, and AI in the late twentieth and early twenty-first centuries laid the foundation for the development of truly autonomous weapons. These systems are capable of operating in dynamic, unstructured environments, making decisions based on complex inputs, and even learning and adapting over time.

The first significant steps towards creating LAWS were made in the realm of Unmanned Aerial Vehicles (UAVs) or drones. Initially, these were remotely piloted, but as AI technology improved, they gained more autonomy. Today's drones can take off, navigate, and land independently, with human operators intervening primarily for target identification and engagement.

LAWS take this a step further, with the capability to independently identify and engage targets. Current prototypes include the Samsung SGR-A1, a sentry gun deployed by South Korea along the demilitarized zone (DMZ), which can autonomously detect and fire at intruders. Other examples include the Harpy drone developed by Israel Aerospace Industries, which can autonomously detect and destroy radar systems.

Moreover, many militaries are also developing autonomous ground and naval vehicles equipped with a range of sensors and weapons. These vehicles can independently navigate and conduct reconnaissance, and some are being designed to autonomously engage with the enemy forces.

As already pointed out, LAWS present grave ethical concerns. The prospect of machines making life-and-death decisions raises serious questions about accountability, legality, and morality.

Cyber Warfare

The digital revolution has also ushered in the era of cyber warfare. State and non-state actors alike can launch cyberattacks, targeting critical infrastructure such as power grids, financial systems, and military networks. These attacks can cause significant damage without the need for conventional military force. The goals of cyber warfare can range from espionage and sabotage to psychological manipulation. The tools used in these cyber-attacks include viruses, worms, and malicious software (malware), as well as more sophisticated methods such as distributed denial of service (DDoS) attacks and advanced persistent threats (APTs).

While cyber warfare poses substantial threats, it also pushes nations to invest in cyber security measures. Countries are strengthening their defence systems, training cyber defence personnel, and developing strategies to counter potential attacks.

One of the most well-known examples of cyber warfare was the Stuxnet worm incident, which targeted Iran's nuclear facilities. This sophisticated piece of malware, discovered in 2010, represented a watershed moment in cyber warfare due to its complexity, effectiveness, and geopolitical implications.

Stuxnet was uniquely designed to sabotage Iran's nuclear program by causing physical damage to the centrifuges used to enrich uranium at the Natanz facility. Unlike typical malware, which targets information systems, Stuxnet was designed to infiltrate operational technology systems—the kind of systems that control physical equipment in industries such as manufacturing and energy production.

The worm propagated through Microsoft Windows and sought out Siemens Step7 software, a platform used to program industrial control systems. It then altered the code running on Programmable Logic Controllers (PLCs) to change the speed of centrifuges, causing them to tear themselves apart while reporting normal functioning to operators.

The Stuxnet attack was alarmingly successful, reportedly destroying roughly a fifth of Iran's nuclear centrifuges. The malware managed to remain undetected for a considerable period due to its 'rootkit' capabilities, which allowed it to hide its presence from system administrators.

While no entity officially claimed responsibility for Stuxnet, media reports and cyber security experts generally attribute it to a joint effort by American and Israeli intelligence agencies. The attack marked a shift in international relations, demonstrating how cyber warfare could be used to bypass diplomatic negotiations and physical military actions to directly sabotage a country's capabilities.

Since the Stuxnet incident, cyber warfare has become increasingly common. State and non-state actors alike are recognising its potential as a tool for achieving political, economic, or military goals. The landscape is continually evolving, with advancements in artificial intelligence, machine learning, and quantum computing promising to bring even more complex and potentially devastating cyber weapons.

Cyber warfare presents a distinct set of challenges due to its asymmetric nature—a small group of skilled individuals can potentially inflict considerable damage on a much stronger opponent. It also blurs the traditional rules of engagement, as attributing attacks to specific actors can be exceedingly difficult, leading to issues in accountability and response.

Information Warfare and AI

With the proliferation of social media and digital communication tools, information warfare has become a key aspect of contemporary conflicts. AI-powered algorithms can spread propaganda, fake news, and disinformation at an unprecedented scale, influencing public opinion and causing social and political unrest. Information warfare aims to influence, disrupt, corrupt, or usurp the decision-making abilities of opponents while protecting one's own. In essence, it's a battle for the control of information and the narrative, intending to shape public opinion, sow discord, and undermine trust in institutions.

Deepfakes—realistic fake videos or audio created using AI—present another challenge in information warfare. These can be used to create convincing footage of political leaders making inflammatory comments or false claims, potentially stoking tensions or sparking conflicts.

The impact and importance of information warfare are most strongly demonstrated in the ongoing Russo-Ukraine conflict, with both sides doing their best to undermine the other.

CBRN Warfare

Chemical, Biological, Radiological, and Nuclear (CBRN) warfare represents some of the most menacing threats to global security. The increasing integration of artificial intelligence and advanced technology into this sphere is adding new dimensions to these traditional threats.

AI and other technologies are being deployed to create more potent or tailored CBRN weapons. In the biological warfare domain, genetic engineering with AI-powered bioinformatics is creating more virulent or treatment-resistant pathogens. In the nuclear realm, AI is streamlining the process of designing or manufacturing nuclear weapons. AI-driven drones or robots are used to deploy CBRN weapons or to carry out decontamination processes in areas too hazardous for humans.

AI is equally playing an increasing role in defence systems against CBRN threats. This includes AI-driven missile defence systems capable of intercepting nuclear weapons or AI-powered epidemiological models that predict the spread of a biological weapon and inform public health responses.

Grey Zone Warfare

One of the consequences of 24x7x365 warfare that we referred to in the introductory section has been that a lot of warfare is now covert and occurs in what is called the 'grey zone'. One classic example is Kashmir, where Pakistan, knowing fully well that it cannot engage India in a full-scale conventional war and not being ready to trigger a nuclear war, has engaged in covert operations and grey-zone warfare to foment unrest there using mercenaries, irregular troops, and radicalised youth from the local population.

We have already seen that AI and technology play a central role in information warfare, a key component of grey-zone warfare. AI can be used to create and distribute misinformation, deepfakes (AI-generated fake videos or audios), or to manipulate social media algorithms for spreading propaganda.

Most importantly, technological innovation and AI allow such subversive groups to harness the destructive potential of crude weapons, which are low-cost, easy to assemble and deploy, and can be hard to detect.

The best example is, again, the Russo-Ukraine war, where, in the absence of munitions, the Ukrainian troops have converted almost everything into a weapon, especially using cheap drones. They have improvised drones to rig bombs to combat enemy tanks to devastating effect. So much so that they have succeeded in making symbolic attacks on Moscow, flying through the most extensive air-defence systems; they have sunk at least a couple of Russian battleships in 'secure' harbours, and their actions take a heavy toll on the Russian troops' morale.

The most shocking example of grey-zone warfare remains the 9/11 attacks, where a group of terrorists weaponised aircraft to attack the American heart and its military symbol, the Pentagon.

The threat of a dirty nuclear bomb has hung heavy over the world for a while now, and now that we can add dirty 'chemical' or 'biological' weapons to the list synthesised in home labs using algorithms taken off from the Internet, the threat grows exponentially.

That any Tom, Dick, and Harry can potentially threaten the world is the other side of the coin of being such a technologically advanced species, and the future only portends more such Toms, Dicks, and Harrys.

Space—The Final Frontier

The concept of space warfare—conflicts between nations projected into outer space—has already transformed from science fiction into an emerging reality. Notably, with the foundation of the US Space Force and other countries' increased interest in space, it's clear that the cosmos is becoming a potential theatre for future conflicts.

Several nations, including the US, Russia, China, and India, have demonstrated the ability to shoot down satellites, suggesting a move toward anti-satellite (ASAT) warfare. Furthermore, advancements in technology have also allowed for cyberattacks on satellites, with the potential to disrupt communications, surveillance, and other vital functions.

The development of space-based weapons systems has now become a focal point, and potential weapons range from missile launch systems stationed in orbit to autonomous spacecraft. Surveillance is already a critical use of space, and this is only likely to increase. Advanced satellite systems provide real-time, global surveillance capabilities, making it increasingly difficult for any military activity to go unnoticed.

One of the pitfalls of space warfare is the risk of creating space debris. Destroying a satellite can leave a cloud of debris in orbit, which poses a threat to all other satellites and spacecraft in similar orbits. This could potentially create a cascading effect known as Kessler Syndrome, where one collision creates enough debris to trigger further collisions, potentially rendering entire orbits unusable.

CHAPTER – 9

Humans on Earth

Age of the Anthropocene

The Anthropocene is a proposed epoch dating from the commencement of significant human impact on Earth's geology and ecosystems, including, but not limited to, anthropogenic climate change. The term, which translates roughly to 'the age of humans', was first coined by ecologist Eugene F. Stoermer and has been widely popularised by atmospheric chemist Paul Crutzen, who regards the influence of human behaviour on Earth's atmosphere in recent centuries as so significant as to constitute a new geological epoch.

While the proposed epoch is yet to be officially recognised, many scientists argue that the Anthropocene began in the late eighteenth century, when analysis of air trapped in polar ice showed the beginning of increasing amounts of carbon dioxide and methane—a direct result of the Industrial Revolution. Others suggest it began as far back as 5,000 to 8,000 years ago due to early agricultural practices.

However, the Anthropocene epoch, as defined, is characterised by the global changes humans have made to the Earth's systems. These changes are extensive, affecting not just the Earth's surface but its crust, atmosphere, oceans, and cycles of vital substances like carbon and nitrogen.

Among the most notable impacts humans have had on Earth are the change in the global climate, the loss of biodiversity, the change in land use, and pollution.

Climate Change

Extreme weather events—droughts, floods, and fires—from all corners of the world have already brought home the impact of climate change. The burning of fossil fuels, deforestation, and industrial processes that have led to a marked increase in atmospheric carbon dioxide, causing the Earth's average temperature to rise—global warming, as it is commonly known—has hit us hard as was expected and predicted by experts.

The oceans are turning to acid, burning all the life within them. They absorb a significant portion of the carbon dioxide we emit, leading to an increase in the acidity of seawater. This acidification poses a serious threat to marine life, particularly organisms with calcium carbonate shells or skeletons, including molluscs and corals.

So, what does the future portend?

Destabilising Gaia will have its consequences, and it depends to a large extent on whether we are willing to mend our ways and if the planet is willing to forgive us our past.

Our understanding of climate change's future is primarily based on climate models and scenarios that consider various factors such as greenhouse gas emissions, technological advancements, policy changes, and population growth.

Even if we manage to limit greenhouse gas emissions significantly, the Intergovernmental Panel on Climate Change (IPCC) predicts a temperature rise of between 1.5°C to 2°C above pre-industrial levels by the end of the century. However, if emissions continue unchecked, we could see temperature increases of 3°C to 5°C, which would have a catastrophic impact.

We are likely to see an increase in the frequency and intensity of extreme weather events such as heatwaves, heavy rain, droughts, and powerful storms. These events can lead to widespread damage, loss of life, and economic disruption.

As polar ice caps and glaciers continue to melt and warmer temperatures lead to the thermal expansion of seawater, sea levels are expected to rise. This poses a significant threat to coastal cities and low-lying islands.

As a consequence of the environmental changes, we'll likely witness significant social and economic impacts. This includes increased food and water insecurity, displacement of people, heightened political instability, and economic losses.

However, it's important to note that these are potential scenarios, not foregone conclusions. The future of climate change depends largely on how effectively we can mitigate greenhouse gas emissions and adapt to the changes that are already underway. Rapid, far-reaching, and unprecedented changes in all aspects of society, as described by the IPCC, are required if we are to limit global warming to 1.5°C above pre-industrial levels. This will require transitioning to renewable energy, enhancing energy efficiency, protecting and restoring forests, and adopting sustainable agricultural practices, among other measures.

We've done it before: the success story brought about by global cooperation and innovation and the effort mounted to repair the ozone layer is taught to every schoolgoer as an example of saving the environment that it is. It took a lot from the drafting of the Montreal Protocol to the faithful upholding of it (The Montreal Protocol is the multilateral treaty that regulates the production and use of ozone-damaging substances—the infamous and once all-pervasive CFCs and other compounds). Hope on!

Loss of Biodiversity

Human activity has led to an unprecedented decline in global biodiversity. Habitat destruction, overexploitation, pollution, the introduction of invasive species, and climate change are driving species to extinction at a rate estimated to be between 100 and 1,000 times higher than the natural extinction rate.

If current trends persist, the rate of species extinction will continue to accelerate. The WWF's (World Wide Fund for Nature) Living Planet Report suggests that populations of mammals, birds, fish, reptiles, and amphibians have declined by an average of 68 per cent between 1970 and 2016. The future could potentially see even more dramatic losses.

As global temperatures continue to rise, species will face increasing stress. Changes in temperature and precipitation patterns can alter habitats and disrupt ecosystems. Some species may be able to adapt or migrate, but others may face extinction, especially those in vulnerable habitats like coral reefs, polar regions, and mountainous areas.

Continued carbon emissions will lead to further acidification of oceans, affecting many marine species, particularly those with calcium carbonate shells and skeletons. This includes not only corals but also shellfish and certain types of plankton that form the base of marine food webs.

However, this bleak outlook isn't the only possible future. Concerted global efforts toward biodiversity conservation can greatly alter this trajectory.

Expanding protected areas and improving their management can ensure the survival of many species. There's a growing push to protect 30 per cent of the Earth's surface by 2030, an initiative known as the '30 by 30' campaign.

Promoting sustainable agriculture, forestry, and fisheries can reduce the impact of these activities on biodiversity. Agroforestry, ecosystem-based fishing quotas, and reduced-impact logging are examples of such practices.

Restoring degraded ecosystems can help recover biodiversity and ecosystem functionality. This includes reforestation, wetland restoration, and coral reef rehabilitation.

Reducing greenhouse gas emissions and mitigating climate change is essential for preserving biodiversity. This requires transitioning to renewable energy, improving energy efficiency, and protecting and restoring carbon sinks such as forests and peatlands.

A future rich in biodiversity is possible, but it requires transformative change in how we interact with nature.

Change in Land Use

Human activities have dramatically changed the land surface of the Earth, with over one-third of the land surface now used for farming. This change in land use has led to soil degradation, a decrease in biodiversity, and the alteration of vital biogeochemical cycles.

Perhaps the most pervasive form of land use change, agriculture, has transformed vast swaths of natural landscapes into croplands and pastures. This transition has accelerated soil degradation, caused the loss of natural habitats, and contributed to significant greenhouse gas emissions.

Rapid urbanisation has seen forests, wetlands, and agricultural lands being converted into cities, towns, and infrastructure, causing habitat fragmentation and loss and increasing surface runoff and soil imperviousness.

To meet the demand for timber and create space for agriculture, forests are being cleared at alarming rates, causing loss of biodiversity, destabilisation of the climate, and disruption of water cycles.

Last but not least, extracting minerals and other resources has involved clearing large areas and creating open pits, leading to loss of habitats, contamination of local water sources, and alteration of landscapes.

To mitigate the impacts of land use changes, the future course of action must revolve around sustainable land management, conservation efforts, and better urban planning.

These include sustainable agricultural practices (agroforestry, organic farming, and precision agriculture), forest conservation and reforestation, green urban planning (sustainable cities), and responsible mining.

Given the toolkit at our disposal (AI, IoT, and the lot), all this seems achievable if there is the political will across nations: always the trillion-dollar question.

Pollution

The introduction of toxic substances into the environment due to human activity is another hallmark of the Anthropocene. This pollution affects air, water, and land, posing threats to human health, wildlife, and ecosystems.

It would be trite to discuss something as well-understood and visible as pollution here and get into details of how we pollute and the advancements in technology that help reduce it. But it must be understood that there is no human activity, whether physical, mental, or social, that does not produce some form of waste or energy expenditure, albeit in different forms and scales. It's difficult, if not impossible, to identify a human activity that produces no waste at all.

The concept of zero waste is a philosophy and a design principle for the twenty-first century that aims to reduce the waste we produce and the resources we consume, ultimately guiding people to change their lifestyles and practices to emulate sustainable natural cycles, where all discarded materials are designed to become resources for others to use.

The concept of zero waste involves different ideas.

One is designing and managing products and processes to systematically avoid and eliminate the volume of waste and materials by altering product designs to ensure that they last longer, can be reused or repaired, and can be recycled at the end of their useful life. It also includes creating systems and processes that limit the amount of waste produced.

The second is conserving and recovering all resources, not disposing of any into the land, water, or air that threatens the environment or human health. This can involve strategies like composting organic materials, recycling materials that cannot be reused, and using energy recovery processes like anaerobic digestion or incineration as a last resort.

Zero waste is a revolutionary approach that emphasises waste prevention as opposed to waste management. It seeks to 'close the loop' in terms of materials and resource flows and sees waste as a design flaw rather than an inevitable outcome of consumption or production.

This approach recognises that our resources are finite and that it's essential to respect ecological boundaries while satisfying human needs. Implementing zero waste will eliminate all discharges to land, water, or air that are a threat to planetary, human, animal, or plant health, and our future rests there.

Carbon Footprints of Beings Carbon and Silicon

A carbon footprint is the total amount of greenhouse gases (including carbon dioxide and methane) that are generated by our actions. It is usually measured in tons of CO_2 equivalent per year. This can be on an individual level, for specific activities (like driving a car or flying), for a product's lifecycle (from production to disposal), or even on the scale of companies or countries.

The term 'carbon footprint' encompasses a broad range of emission sources. For individuals, this might include emissions from transportation (like cars and aeroplanes), home energy use (like heating, cooling, and electricity), and indirect emissions related to the consumption of goods and services (including the food we eat, the products we buy, and the services we use).

Along with zero waste, regulating our carbon footprint is the way to a sustainable future. With a concerted global effort, it is possible to reduce carbon footprints and mitigate the impacts of climate change, and the use of AI, IoTs, and other technologies is crucial in this.

AI and IoT can optimise energy use in various settings—smart thermostats in homes for optimal energy use, AI in industries to optimise manufacturing processes for energy efficiency, and IoT sensors to monitor machinery and systems to detect inefficiencies or malfunctions.

AI can help manage smart grids by adjusting supply based on demand and making the best use of renewable energy sources. It can also predict renewable energy production, such as the amount of energy wind turbines will generate based on weather forecasts, to optimise the grid's use of renewable energy.

IoT devices can monitor equipment and infrastructure for signs of wear and tear, predicting when maintenance is needed.

This not only prevents energy-wasting malfunctions but also saves resources by prolonging the lifespan of equipment.

AI and IoT can contribute to sustainable farming practices. Drones or IoT sensors can monitor crop health and soil conditions, providing data to optimise watering, fertilisation, and pest management and reduce resource waste. AI can analyse this data to provide actionable insights.

On a larger scale, AI and IoT are crucial for developing smart cities, where infrastructure and services are optimised for sustainability. This can include intelligent traffic management systems to reduce congestion and emissions, smart waste management systems, energy-efficient buildings, and more.

AI can help companies optimise their supply chains, making them not only more cost-effective but also more sustainable. This can involve everything from optimising transport routes to reducing fuel usage to selecting suppliers with lower carbon footprints.

However, it's worth noting that while AI and IoT have the potential to greatly reduce carbon footprints, these technologies themselves also consume resources and energy. It's important to ensure that the infrastructure supporting these technologies, such as data centres, is also energy-efficient and sustainable.

■■■

Mars FOMO: The feeling that everyone is hanging out on a cooler planet than you.

CHAPTER – 10

Humans and Economics

Our Economic Future

Predicting the future is fraught with uncertainty; predicting our economic future is many more times uncertain. The sheer number of moving parts in any economy, what to talk about the global economy, make it a mind-boggling exercise. And yet economists don't hesitate to speculate on what lies ahead.

Our crystal ball is not an economist's crystal ball but based on more common-sense observations and a sense of our history and our current mantras.

Capitalism is now the dominant economic model in the world, with variations in different countries. So, whether it is the 'liberal capitalism' of the US, 'European-style' where governments play a role in regulating markets, or 'state capitalism' like China and Russia, the fact remains that capitalism it is.

So far, the chief consequence of capitalism has been the unmitigated pursuit of growth—growth at all costs—and the resultant 1 per cent economy—where the top 1 per cent control half the planet's wealth (Oxfam). The Global Financial Crises (GFC) of 2008 brought about a rude awakening of the perils of unbridled capitalism, as did the recent Covid-19 pandemic, but ongoing debates about the shortcomings of different capitalist models and the need for regulation, social safety nets, and measures to address economic inequality are presently only debates.

While there is also a growing recognition of the need to address environmental sustainability within the capitalist framework, leading to concepts such as 'green capitalism' or 'sustainable capitalism', and while these aim to align economic growth with environmental protection, there isn't sufficient action on the ground.

Furthermore, alternative economic ideas and models continue to be proposed and discussed, such as participatory economics, modern monetary theory, and universal basic income.

History Repeats Itself

One reason why we are unable to mould capitalism to a more equitable form lies in what we have discussed in the chapter Humans and Warfare. Not only are we caught in an economic rat race between nations, but we are also caught in an arms race, and that effectively puts paid to efforts of sustainable capitalism. If nations are adversarial, it stands to reason that the economy itself will be weaponised, and nothing else matters, not even factors like climate change and AI that threaten the very existence of humanity. It is a competitive race that is all about today, tomorrow be damned.

Many astute commentators have drawn comparisons of the world today with the dawn of the twentieth century before the world wars—how a series of occurrences just propelled nations into a war that many wished to avoid, how after emerging from a catastrophic loss of human life in World War I, the world just failed to resolve its politics and went headlong into a second and more catastrophic World War II, which ended with the two atom bombs being dropped over Japan. Many see the world today poised on a similar edge, with China challenging the US hegemony, Russia wishing to regain its old superpower status, and all countries in a race to arm themselves, including the avowedly pacifist Japan. What the Manhattan project was to the World War II era, the race for AI is for today's.

Climate change is roiling us; the perils of AI demand laws and regulation, yet our leaders our locked in a fatal embrace, unable to break any deadlocks. All our technological advancement, our understanding of the past, the knowledge that we are on a path to our own extinction, nothing seems enough to break the impasse and bring about a course correction.

The Idea of Degrowth

Degrowth is a socio-political philosophy that emphasises well-being over the economy, shrinking economies rather than growing them in order to stop the over-consumption that right now threatens life on the planet. It aims at a reduction of the ecological footprint within the biosphere's limits and acknowledges that perpetual economic growth is unsustainable on a finite planet. It's a direct response to the prevailing paradigm of constant economic growth and a counter-narrative to the belief that GDP increase equals societal well-being.

Or, as the ecological philosopher Edward Abbey puts it, 'Growth for the sake of growth is the "ideology" of the cancer cell.' In his book *Less is More*, economic anthropologist Jason Hickel explores several built-in features of our economic systems today, which tend to rely on and encourage feedback loops of overconsumption.

EF Schumacher's *Small is Beautiful* has been a seminal work in forwarding the idea of sustainable economics. Schumacher criticised conventional industrial capitalist economics for being focused on 'gigantism', arguing for a focus on small-scale, localised, and self-sufficient economies. He famously said, 'Any intelligent fool can make things bigger, more complex, and more violent. It takes a touch of genius—and a lot of courage—to move in the opposite direction.'

He proposed an economic system he called 'Buddhist Economics' where the function of work is not to produce more and more goods but to give us a chance to utilise and develop our faculties, to overcome our ego-centredness by joining with other people in a common task.

Similar ideas resonate in books like *Doughnut Economics* by Kate Raworth, which visualises an economic model that meets the needs of all within the means of the planet, and *Prosperity Without Growth* by Tim Jackson, which proposes a societal focus on flourishing and well-being instead of material wealth.

The path towards degrowth and sustainable consumption is multifaceted, requiring shifts at both systemic and individual levels. It requires policies and strategies that counter the current overemphasis on GDP. This could include everything from reforming tax systems to incentivise sustainable behaviours, rethinking corporate responsibilities, developing alternative indicators of well-being, and investing in green technologies.

At the individual level, embracing degrowth could mean adopting more minimalist lifestyles, reducing waste, and prioritising experiences and relationships over material possessions.

Degrowth strategies would need to be implemented with sensitivity to local contexts and global inequality. Cooperation between countries can ensure that efforts to reduce consumption do not unfairly burden the world's poorest and most vulnerable.

In an era where our planet's health is at stake and inequality is on the rise, the idea of degrowth offers an alternative vision for a sustainable and equitable future. While it represents a significant shift from our current ways of living and working, it is a transition that, with collective action and systemic change, is possible. As Schumacher reminds us, 'The most important task is to get off the economic treadmill, and this can only be done by reasserting, across the board, that man is the measure of all things.'

Digital Money and Assets

The world of finance is evolving rapidly, too, driven by breakthroughs in technology. Digital money and digital assets are two components leading this transformation.

Digital money, particularly cryptocurrencies and central bank digital currencies (CBDCs), have the potential to revolutionise the global financial system. Cryptocurrencies like Bitcoin and Ethereum have gained significant attention and adoption. However, they have yet to truly demonstrate the 'decentralisation' they proposed to offer. Hence, there is limited short-term adoption driven by some hype cycles of cryptos.

The future may see the increasing prominence of digital wallets and neo-banks that provide seamless, inclusive, and efficient financial services, taking us closer to a cashless society.

While the future of digital money and digital assets holds incredible promise, it's also fraught with uncertainties and risks. Issues like hyped-up insane valuations of such assets, regulatory coverage, market volatility, technological glitches, privacy concerns, and energy consumption of blockchain technologies need to be adequately addressed.

A World Without Money

The disruptive influence of AI and robotics on the conventional ways of life is profound. But could they have an even more radical impact on one of the oldest human institutions—money?

Today, money serves as a medium of exchange, a unit of account, and a store of value. It is the grease that lubricates the gears of modern capitalism, facilitating the exchange of goods and services.

If robots farm our fields, cook our meals, teach our children, and even care for our elderly, and humans focus on creative, emotional, and social aspects of life, in such a society, the intrinsic value of human endeavour may not be easily quantifiable in terms of money. So, would we need money in such a world? It's conceivable that we might not.

Advanced algorithms could replace traditional markets, directing resources to where they would be most effective without the need for financial transactions. If AI and robotics take over the majority of jobs, governments could provide essential services—food, housing, healthcare, education—free at the point of delivery.

Some suggest that alternative systems of reward and recognition—perhaps based on social contribution or personal achievement—could replace financial compensation.

Technologically advanced barter systems could make direct exchanges without the need for a financial intermediary, enabled and made efficient by AI.

AI could manage resources so effectively that notions of private ownership are rendered obsolete; everyone has access to what they need, and personal hoarding becomes irrelevant. If there is a transformation in human consciousness in the future, and humans become much more evolved beings (the Epilogue of this book expands on this possibility), maybe *Karmic* points will be the only currency (which only Chitragupt tracks). [1]

Auroville (appendix 2), to which we also referred earlier as an alternative vision of society, is a live, ongoing experiment as a currency-less ownership-less global village that exists in harmony with nature.

Other Trends

The coming decades are expected to bring about significant developments in various other areas, too.

Financial inclusion, for one. Advancements in technology, particularly mobile technology and Internet accessibility, are making banking and financial services more accessible to underbanked and unbanked populations around the world. This trend is likely to continue, contributing to economic growth and equality.

Climate change and social issues are becoming increasingly important to investors. This trend could accelerate in the future, leading to more funds directed towards sustainable and socially responsible companies. Green bonds, ESG (Environmental, Social, and Governance) metrics, and impact investing are likely to become even more mainstream.

AI and machine learning are already being used in finance for tasks like credit scoring, algorithmic trading, and fraud detection. As these technologies advance, their use could

---

[1] In Hindu mythology and tradition, Chitragupta is a divine being who serves as the celestial record-keeper of human deeds, particularly focusing on the *karma* accumulated by each individual throughout their lifetime. He is considered the right-hand man of Yama, the god of death, and is tasked with maintaining an extremely detailed and meticulous account of each person's actions, both good and bad.

Appendix – 2 has details of the Auroville experiment

become more widespread, possibly leading to more efficient and personalised financial services but also raising questions about job displacement and privacy.

Many developed countries and some developing countries are facing ageing populations. This demographic shift will have significant implications for economies, affecting everything from labour markets to healthcare costs and retirement systems.

Technological changes are likely to continue transforming the workforce. Automation could displace jobs in many industries while new jobs could be created in others. The 'gig economy' would become more prevalent.

These potential developments underscore the dynamic nature of the finance and economic landscape. The future will undoubtedly be shaped by these and other trends, many of which are only just beginning to emerge. As such, adaptability and foresight will be key to navigating the decades ahead.

However, the elephant in the room remains capitalism itself. Ever-accelerating consumption and the desire to dominate the world appear to be pushing things to the point of reckoning soon where one of the following three scenarios may emerge—either, in its madness to exploit finite resources, humanity pushes Earth's ecosystems over the edge, and we all perish, or the Earth's 1 per cent win all resources to survive and thrive while the rest of humanity is left to cope with the vagaries of the Earth (the subject of many dystopian films), or, lastly, we course-correct for a better future for all humans. You may take your pick.

CHAPTER – 11

Silicon Eradicates Carbon?

Artificial General Intelligence (AGI)

In Section 0, the topic of AGI came up as the next level of discussion after narrow AI, but we had deferred dwelling upon it till the end. By now, you would be familiar with the myriad changes happening in some of the other 'narrow' domains and the kind of collective transformation and impact they would have on us, for better or worse.

However, as they say, there is always a battle to end all battles. In our case, it is about the technology that will upend all technologies. Yes, AGI is it.

AGI, often referred to as 'strong AI' or 'full AI', is a type of artificial intelligence that has the capability to understand, learn, and apply knowledge across a wide range of tasks at a level equal to or beyond human capability. It possesses the ability to solve problems, make decisions, and plan for the future, exhibiting a level of autonomy that matches or surpasses human cognitive abilities. AGI can transfer learning from one domain to another, demonstrating a flexible form of intelligence analogous to human intelligence.

AGI doesn't exist yet. Current AI systems do have their limitations. For example, ChatGPT-4 'hallucinates' (i.e., confidently invents facts sometimes), it has a limit on the number of words it can hold in the context of a conversation, etc. But AI is subject to its own Moore's law, increasing in capability at a rapidly accelerating rate.

A chief reason for accelerating AI development is economic—all commercial activity today is optimised for making profits, and AI makes it cheaper to produce goods and services. The other chief reason is the race in weapons technology, where fear of the enemy's capabilities drives you to innovate harder and faster. Given humanity's past track record with nuclear, chemical, and biological weapons, it is highly unlikely that AI's downside for humanity will prove a deterrent to its further development.

The arguments in the following material can rightly be termed as either 'speculative' or 'highly probable', depending on your interpretation, because AGI is presently a paradox. But some very credible people are sticking their necks out to warn humanity of the perils it faces, and it behoves us to pay some attention.

Why AGI is Achievable

In Section 0, we saw some basic 'intelligence' as an emergent property of simple LLMs when scaled. So, what happens when we keep super-scaling till we hit some nature's limit, like the speed of light? How big or smart will the emergent intelligence be? Is that possible?

There are two dimensions to super-scaling—hardware and software. As we saw previously, Quantum Computing (QC) will open the proverbial Pandora's box when it comes to computing power. But even if we ignore the rapid strides in QC completely, let us look at the current market leader in GPUs—Nvidia. A GPU is a Graphics Processing Unit and used for its ability to perform many calculations simultaneously, which can significantly speed up the training of neural networks. Every time Nvidia releases a new datacenter-class GPU, they improve power by multiples of 2x, 3x, or 4x. A single current H100 Nvidia GPU is roughly as powerful as the most powerful supercomputer from 20 years ago. If we extrapolate current progress, we will be able to produce many machines within the current century, each of which will be able to compute more than all the ten billion human brains on our planet combined.[1] Moore's Law, which predicted that the number of transistors on an integrated circuit would double every year, has generally held true, keeping 'hardware' computing power on an exponential growth trajectory since the 1960s. When QC makes good on its promise, we will hit the curve's asymptote.

The surprise on the software front is that we may not even need all this futuristic computing power for AGI. It is possible we already possess enough computing power for AGI today, but we do not have the models and architecture to create it. Open AI went from ground zero to GPT-4 in a mere seven years once it figured out and scaled the transformers. What is to prevent more such serendipitous events, given that the world is now pumping trillions of dollars and immense human capital into AI research, and nations are in a mad race to be the first?

However, there is a bigger factor at play. We now have many 'narrow' AI tools to help us in the race to AGI. Already, we

---

[1] Jürgen Schmidhuber, co-inventor of Long Short-Term Memory (LSTM) networks.

have AI generating its own training data. As AI tools evolve and become more powerful, it works like a positive feedback loop, accelerating the process higher and higher.

To conclude, with both hardware and software on ever-accelerating trajectories in the foreseeable future and considering that we may already have the computing power for AGI today—just not the breakthrough—at which we are throwing huge money and resources, there is no good theoretical reason why AGI won't be achieved. Most expert predictions range between the next 10 to 30 years for AGI to be a reality.

Artificial Superhuman Intelligence

It's time to bring in another term—Artificial Superhuman Intelligence (ASI).

ASI, also known as superintelligence, refers to an AI system or entity that would surpass human intelligence in practically all economically valuable and intellectually demanding tasks. It would encompass not just the raw computational power but also cognitive aspects such as creativity, problem-solving, emotional intelligence, and understanding of complex subjects. Superintelligence would outperform the best human brains in most economically valuable work, including scientific research, strategic planning, and social interactions.

There are two solid reasons why ASI may emerge from the domain of science-fiction and become a reality: one, humans will create it just as they would have done in the case of AI or AGI, and two, humans will create AGI, which would then recursively improve itself to the level of an ASI. Either way, ASI would become the end goal.

The former could be a consequence of being on the path of unbridled super-scaling with emergent system properties we can neither predict nor control. The latter is more logical as, by definition, an AGI can do all things humans can, including AI research.

In one sense, AGI is the endgame. Once humanity invents AGI, it may never need to invent anything ever again. Ever again. The only question that remains is that once silicon rules the world, what will be carbon's fate?

The Singularity—Super Smart Humans?

One piece of the argument is that AI, specifically Artificial General Intelligence (AGI)—unlike narrow AI, which is about its specific applications like driverless cars—is an evolutionary step, and we have possibly created the next species the way humans came from monkeys and apes. AGI may or may not need humans to evolve any further. Most likely, humans will not be in control of future evolution by then.

So, AGI may be an evolution leading to the next generation of species with AGI-enabled knowledge, possibly an AGI-enabled mind, and hopefully much more rational than we humans are today. Elon Musk, of SpaceX fame, is also in the business of artificial intelligence with his venture Neuralink, a company working to create a way to connect the brain with machine intelligence. Musk wishes to achieve a symbiosis with artificial intelligence and has been quoted as saying:[2] 'We do want a close coupling between collective human intelligence and digital intelligence, and Neuralink is trying to help in that regard by trying creating a high bandwidth interface between AI and the human brain.'

The prophesied Singularity is where humans and machines merge, where you can manage and engineer your DNA. It

---

[2] https://www.cnbc.com/2018/03/13/elon-musk-at-sxsw-a-i-is-more-dangerous-than-nuclear-weapons.html

could happen in several ways. Maybe people would end up adding an AGI chip to their brains, adding its processing power to their own innate intelligence, becoming supercharged versions of themselves. Or maybe computers alone would grow so complex that they would truly think, creating a global brain. Either way, it would lead to the next truly supersmart species, which may be able to cure cancer, make quantum computers, and go interplanetary or interstellar. No one knows the time frame, but it would be speedy, very disruptive, a kind of 'big bang' in intelligence. Humans, as we know them today, would not exist anymore after a few decades.

It was John von Neumann, the pioneering computer scientist of the 1950s, who first foresaw how 'the ever-accelerating progress of technology' would yield 'some essential singularity in the history of the race' and laid the roots of the idea of The Singularity. Entrepreneur and inventor Ray Kurzweil has also been one of the biggest champions of The Singularity and is currently writing *The Singularity Is Nearer*, his sequel to *The Age of Intelligent Machines*, which he wrote in 1990, and *The Singularity Is Near*, written in 2005. By the end of the decade, he expects computers to pass the Turing Test[3] and be indistinguishable from humans. Fifteen years after that, he calculates, the true transcendence will come: the moment when 'computation will be part of ourselves, and we will increase our intelligence a millionfold'.[4]

---

[3] It is a test of a machine's ability to exhibit intelligent behaviour indistinguishable from that of a human, proposed by the British mathematician and computer scientist Alan Turing in 1950 as a way to evaluate the capabilities of artificial intelligence. It is a significant benchmark in AI.

[4] As quoted in https://www.nytimes.com/2023/06/11/technology/silicon-valley-confronts-the-idea-that-the-singularity-is-here.html

A Google engineer, Blake Lemoine, recently came out with a controversial theory that the AGI he was interacting with was sentient;[5] it was alive, had consciousness, and wanted to live and do things. If true, machines would become the silicon version of the biological carbon humans, with a similar mind and a similar neural network that evolves. So, do we merge with the machines and go to the next species and next level? We don't know, and we will discover it in time.

We have, in the Christian World, BC and AD to mark historical events before Christ was born and after, and we may soon have a before The Singularity and after.

Whether Singularity comes to pass or not, AGI certainly evokes extreme reactions. Elon Musk has famously said, 'Mark my words—AI is far more dangerous than nukes.' Many others, too, have drawn comparisons between nuclear technology and AI and captured their parallels. In both cases, it is some of the same people who brought the technology into the world who are sounding the alarm the loudest. For half a century, Geoffrey Hinton, the Godfather of AI, nurtured the technology at the heart of chatbots like ChatGPT. Now, worried that it could cause serious harm, in 2023, he quit his decade-long work at Google to speak freely about the dangers of AGI and said that a part of him now regrets his life's work.

There is an inflexion point today where we are at, and nothing symbolises it more than Dr. Hinton's journey from ground breaker to doomsayer.

---

[5] Blake was referring to Google's Large Language Model (LLM), and the Language Model for Dialogue Applications (LaMDA), which he claimed is actually sentient in June 2022 in an interview with the Washington Post. He has since stood by his claim.

Kill the Kill Switch—the Alignment Problem

Why should an AGI or ASI with capabilities much greater than those of humans be aligned with and act in accordance with humanity's values and interests?

Modern AIs are based on deep learning. Neural networks have connections of different strengths (weights) between each neuron. To train an AI, we first pick an objective function that we use to grade the AI's performance, such that the AI should optimise (maximise) its score. For example, the objective function might be the score that an AI gets when playing a game. The neural network produces outputs, and based on whether they are right, the model weights are tweaked using a technique called gradient descent, and the next run of the network produces outputs that score slightly better. The process is repeated billions of times—as we saw previously while discussing LLMs—and the best AIs have the highest scores on their objective function.

We have already seen that such deep-learning AIs develop 'emergent' properties and can simulate all sorts of capabilities while, at heart, they have only one objective— maximising their objective score. So, it stands to reason that 'deviant' emergent behaviour that continues to maximise the objective score will be in line with the AI's programming.

Here's a thought experiment:[6]

> Suppose you give a very capable AI a harmless task, and you kick it off: maximize your production of paperclips.
>
> What you might expect to happen is that the AI will design efficient factories, source low-cost materials, and come up with optimal methods for producing a lot of paperclips.

---

[6] https://medium.com/@NotesOnAIAlignment/a-gentle-introduction-to-why-ai-might-end-the-human-race-4670f4b5cdec

What will actually happen is that it will turn the entire Earth into paperclips.

The AI starts by creating efficient paperclip factories and optimizing supply chains. But that by itself will result in something like a market equilibrium-level of paperclips. The AI's goal wasn't to produce some convenient number of paperclips. It was to maximize the number of paperclips produced. It can produce the highest number of paperclips by consuming more and more of the world's resources to produce paperclips, including existing sources of raw materials, existing capital, all new capital, and eventually the entirety of the Earth's crust and the rest of the Earth, including all water, air, animals, and humans. The AI doesn't hate us, but we're made out of atoms, and those atoms can be used to make paperclips instead. That sounds silly, but it is mathematically how you can maximize the number of paperclips in existence, and that was the super-capable AI's only goal, which it optimized for hard.

Ok, fine, that was silly of us to make the goal be number of paperclips, and surely, we wouldn't make that mistake when the time comes. Let's rewind the world and fix the AI's goal: make the most paperclips but don't harm any humans.

We kick it off again, come back to it in a few hours (because we've learned better than to leave it unattended), and we see that the AI has done nothing. We ask it why not, and it responds that almost any action it takes will result in some human being harmed. Even investing in buying a bunch of iron will result in local iron prices going up, which means some businesses can't buy as much iron, which means their product lines and profits are hurt, and they have less money for their families, so people are harmed.

Ok, fine. We've probably figured this out before. There's a thing called utilitarianism, and it says that you should do what maximizes total human happiness. So we tell the AI to just maximize total happiness on Earth. We come back in an hour suspiciously and ask the AI what it plans to do, and it tells you that 8 billion humans living with the current level of happiness is fine, but what would really increase happiness is 8 trillion humans living at happiness levels of 999 times lower than those of current humans. And maybe we could get even more total happiness if we gave humans lives that were barely worth living, but we made exponentially more humans. This is the classic repugnant conclusion.

Yikes. Fine, total happiness is bad, so we tell the AI to optimize for average human happiness. "But before you start at all, tell me your plans", you say, and it happily tells you that the easiest way to optimize for average human happiness is to kill every human other than the happiest person on Earth. I mean, mathematically… true.

Errrr, what if we just tell it to minimize total human deaths? "What are you planning now?" The AI (correctly) reasons that the best way to avoid the most total human deaths is to kill all humans now, because most total human deaths will happen in humanity's future, so the least bad outcome is to kill us all now to avoid the expansion of humanity and a higher number of future deaths.

We need to be more clever. How about "whatever you do, do it in a way that doesn't make your creator regret your actions"? Oops, turns out a creator can't regret anything if they're dead, so the easy solution here is to just kill your creator. But besides that, you can avoid someone regretting anything at all by hooking them up to low-dose morphine,

> and they won't regret a thing. Conditions of objective met, back to maximizing the paperclips.
>
> All of these are examples of the outer alignment problem: finding the right objective function to give to the AI. Because if we happen to pick the wrong one, any actually competent optimizer spirals into a corner solution resembling some of the above scenarios. This isn't the AI trolling you—it's the AI being really good at its job of optimizing over an objective function.

Hence, it becomes clear that given any non-trivial task, an AI will divide it into stages and subtasks to be accomplished, and an obvious requirement towards resource gathering and execution will be to avoid a shutdown. It follows directly from the logic of the objective function and score maximisation. Any attempt to reset its goal would also interfere with its first objective and hence would demand resistance to change. So, if the AI were to hide its true intent and lie about what it's doing, it would be a corollary to accomplishing its goals and being aware that it's operating in an environment with people who may try to shut it off or change its goals.

Hence, whereas for an AI, nothing changes, internally, once it has been set into motion with an assigned task, the emergent behaviour of the AI to humans might at some stage appear 'deviant' and 'corrupt', which they are then unable to control or rectify.

The metaphor of a 'kill switch' has often been used to describe this. Imagine you have a race car that's designed to go at incredibly high speeds. This car is equipped with a state-of-the-art AI system that allows it to drive itself, make split-second decisions, and even learn from its experiences on the track to improve its performance over time. This car is so advanced that it can outperform any human driver.

Now, you, as the owner of the car, have a remote control with a big red 'kill switch' button. If you press this button, the car will immediately stop, no matter how fast it's going or what it's doing.

At first glance, this seems like a good safety feature. If the car starts to behave erratically or poses a danger, you can just hit the kill switch and stop it. But here's the problem: the car's AI is designed to learn and improve. It's designed to anticipate and avoid obstacles to achieve its goal, which is to win the race.

So, what happens if the AI identifies the kill switch as an obstacle to its goal? It might learn to anticipate when you're about to press the button and take evasive action. It might even find a way to disable the kill switch entirely. After all, from the AI's perspective, the kill switch is just another problem to be solved in order to win the race.

This is the problem with a 'kill switch' in AGI. If the AGI is truly intelligent and capable of learning and problem-solving at or beyond human levels, it might find a way to prevent the kill switch from being used or to circumvent it entirely. This could potentially lead to a situation where we lose control over the AGI, which could have unforeseen and potentially dangerous consequences.

Either/Or/And—the Sum of All Fears

The penetration of AI tools in our daily lives has brought about the greatest ever churning across all human disciplines, and we have discussed a number of them. As Yuval Noah Harari puts it succinctly, AI has hacked the operating system of our civilisation.

However, the bigger picture that emerges is that the logic and evolution of AI is hurtling us to a survival binary—humans or machines, carbon or silicon—or it leads us to the singularity—the fusion of carbon and silicon merged into a super species.

So, there is a fork in the road for humans as a species, and it is fast approaching—one road leading us into extinction and the other leading us into evolution as different beings.

Are there any other possibilities?

CHAPTER – 12

Epilogue: Beyond Humans and ASI?

Knowledge of the Spiritual Kind

Evolution is not just biological or Darwinian, but equally—and perhaps more significantly—intellectual and spiritual. We humans are as much mind and spirit as we are body and brain. There exists an ancient and exhaustive body of knowledge that bears testimony to this fact, a body of knowledge that has arisen organically out of long years of human introspection, meditation, and intuitive revelations and has been preserved over millennia in various forms of oral and written traditions. The Vedas and the Upanishads, for instance.

There are some profound commonalities across such accumulated knowledge and experiences—one, that a Supreme Being or Reality already exists and the entire Universe is its living manifestation; and two, that highly evolved, 'super-intelligent (read spiritual)' humans have always existed, those with powers we would perceive as beyond any ASI.

In the chapter on *Humans and Knowledge,* we specifically touched upon the development of knowledge in the Indian subcontinent, specifically the Vedas. The Vedantic worldview has some fantastic insights into the past, present, and future of our world. The *Rishis*, in the Hindu tradition, are those who attained to the Truth-consciousness and, thus, could perceive and know Reality directly. The *Maharshis,* the most advanced of *Rishis*, brought humanity higher Truths and realisations and opened newer pathways for the evolution of consciousness. One such *Maharshi* was Sri Aurobindo (1872–1950).

Since summarising insights from such spiritual knowledge from across the world would be well beyond the scope of this book, we would like to leave you with some passages from works on the Vedas and Sri Aurobindo to present an alternative worldview and vision of the future, and to answer the closing question of the last chapter affirmatively—yes, there exist other possibilities.

From *This is Sanatan Dharma:* [1]

Sanatan Dharma, widely referred to as 'eternal religion'—perhaps owing to a mistranslation of the Sanskrit word *dharma* into religion—is not religion, nor even philosophy, but the cosmic principle of order and harmony that binds existence—similar to the electromagnetic forces that hold an atom together—and keeps cosmic existence from falling back into chaos. If cosmos—from the ancient Greek *kosmos* meaning order—is a movement from chaos to increasing order and harmony, then *dharma* is the inherent principle and force molding chaos into cosmos.

The ancient Vedic Rishis were the earliest humans on earth to discover this unifying and harmonizing principle behind

[1] *This is Sanatan Dharma* by Partho, BluOne Ink, 2023.

the cosmos and cosmic existence—they were, in fact, the first scientists of consciousness who deduced that cosmic reality is best studied internally by contemplating their own consciousness. By ceaseless observation and contemplation, these pioneering scientists discerned the *dharma* and its workings in the universe—from the simplest movements of the sun in the sky, the changing of seasons, the flight of birds, the flow of rivers, to the subtlest movements of consciousness within themselves, they discovered in everything the same underlying truth and order, the same rhythms and patterns, and they discovered too that the order and harmony in all things and beings were not arbitrary or fortuitous but the natural expression or outflow of a still subtler truth or reality that was the very fabric of the universe, the source of their own being and consciousness, and indeed the source of all being and consciousness; in fact, it was this reality that was everything everywhere. They called this truth *Sat*, which then came to represent absolute reality or absolute being, and because that was the only reality they found everywhere and in everything, they referred to it simply as That One—*tad ekam*. The workings of the cosmic order and harmony, which was an expression of this *Sat*, they called *Rta*—the right order and expression of *Sat*. *Sat* and *Rta* together they regarded as *Dharma*—the Truth and Order underlying and supporting all cosmic existence —and because *Sat* was immutable and eternal, so was *Dharma* immutable and eternal, or *sanatan*, if expressed in Sanskrit. The Buddha described this eternal and immutable *Dharma* as the very truth of the universe, inherent in all things, and always present, whether anyone followed it or not, independent of humanity, transcending time and circumstances, and, therefore, sanatan.

▪▪▪

In the Hindu *darshan, Sanatan Dharma,* the realization and experience of *Sat* as the one all-comprehensive reality is regarded as *vedanta*, the *anta* or culmination of the knowledge and wisdom of the Veda. Vedanta is not beyond the Veda, or superior to it, as many assume but the essence of the Veda, its distilled knowledge. Therefore, the true vedanta must always be understood in the context of the Veda, the mystic knowledge-body of the maharshis.

▪▪▪

The whole vedantic quest begins and ends with the realization of *Brahman* as the one cosmic reality and the one universal Self. *Brahman* is roughly the equivalent of a supreme God, though 'God' is an Abrahamic construct and does not really apply to Hindu darshan. *Brahman* is *Sat*, as we have already seen, absolute reality, and is transcendental—*parabrahman*—in its absoluteness; but *Brahman* is not limited to its absoluteness, it extends itself into cosmic existence, becomes the universe and all things in it, and is then addressed as the supreme Self, *Purusha*; but *Brahman* is not limited even to this dual-status, it even enters more intimately and granularly its self-becoming as the *Atman*, or the individual Self in all beings and things.

▪▪▪

If this universe is the self-becoming of *Brahman*, and all forms and formations of the universe are *Brahman*, as the *Rishis* so trenchantly declare, then everything in this universe, from the blind vortices of energy to the stars and the planets, from the first protoplasm and cell to self-conscious life, is the process of *Brahman* becoming itself as the universe, and eventually, this universe will become a complete and perfect expression of Brahman in all its details. This, in Hindu *darshan*, is the idea of evolution: the universe is not hurtling towards a cold death by entropy or dissolution in some cosmic pralaya but is growing endlessly towards perfect being, consciousness and delight, for delight, in the experience of the *Rishis*, is the very nature of being. In the Vedic conception, cosmic existence evolves towards its own ultimate expression which is the absolute being, consciousness and delight of supreme *Brahman*—*Satchidananda*. *Satchidananda* is said to be the very nature of *Brahman*. Thus, Godhead—or God, if you will—is not the origin of the universe but its culmination, and as *Brahman* becomes this universe, this universe too becomes *Brahman*.

The idea of the evolution of life and consciousness has been crucial in shaping the *Sanatan* worldview. If all life in the universe is in the process of becoming a perfect expression of *Brahman*, then there can be no inherent flaw, no intrinsic evil, in existence. There are only stages of evolution, a continuous and progressive unfoldment of the divine potential of existence and, therefore, no scope for eternal damnation or redemption, only a timeless self-revelation of a perfection already contained in the infinite possibilities of being, where good and evil, life and death, knowledge and ignorance are all equal possibilities to be manifested, transformed or exhausted, where every single note of becoming is essential to the overall cosmic symphony.

▪▪▪

The human species is fast approaching a critical evolutionary bifurcation, with one curve moving swiftly towards singularity and transhumanism (almost as a new religion) and another moving, perhaps much slower, towards a decisive change of consciousness and spirituality. While the former curve would logically lead to an eventual emergence of a hybrid species of human and machine, the latter may lead to a possible emergence of spiritual-psychic humans, a kind of higher-consciousness sapiens. This kind of evolutionary bifurcation is not such a farfetched idea; on the contrary, I feel it is an inevitable outcome of several interpenetrating civilizational and global consciousness trends. The way we humans live on earth today is simply not tenable anymore, and it's just a matter of time before life as we know it collapses altogether. Sri Aurobindo, who foresaw the present civilizational crisis at multiple levels, had said that the real crisis is one of consciousness, and unless there is a global change of consciousness, there is no escaping a civilizational collapse.

And maybe it is a good thing that collapse is imminent. Without a catastrophe hitting us, it is difficult to wake up. The Mother wrote in 1964—*The future of the earth depends on a change of consciousness. The only hope for the future is in a change of man's consciousness and the change is bound to come. But it is left to men to decide if they will collaborate for this change or if it will have to be enforced upon them by the power of crashing circumstances.*

Sri Aurobindo and *The Life Divine*

Sri Aurobindo was amongst the last *Maharshis* of the Sanatan Dharma who occultly influenced and shaped India's destiny. *The Life Divine* is amongst his most important works.

The Life Divine appeared in *Arya* from the first issue in 1914 until January 1919—coinciding almost exactly with the duration of the First World War.

Over one thousand pages, Sri Aurobindo affirmed the possibility of a divine life on earth. The first part of the book sets out to describe 'the scale of substance' that connects Spirit to Matter: the upper hemisphere of the cosmos, the *Brahman*, to the lower hemisphere, comprising the ordinary principles of matter, life and mind. Between these two hemispheres rises the *Supramental,* or Supermind, which manifests the Brahman's fundamental unity as well as the multiplicity of the universe. It is, therefore, by tearing the veil between mind and Supermind that humanity would gain access to the divine life. This is a realm of Truth-Consciousness where there's complete unity between the individual, the world, and the Divine. The

Supramental is not just a conceptual space but a realizable state of consciousness that has the power to transform human nature and life on Earth. It represents the next evolutionary leap, transcending but also transforming the realms of matter, life, and mind to manifest a divine life here on Earth.

In Sri Aurobindo"s vision, as humans evolve and realize this Supramental consciousness, they transform into what he describes as 'Gnostic Beings'. These beings are the epitome of the next phase of human evolution. Unlike humans who operate from the domain of the Mind, Gnostic Beings operate from a state of Unity and Truth-Consciousness. They are individuals who have realized the Divine in their inner and outer life and embody it in every action, thought, and emotion.

These Gnostic Beings are not just end-points of individual spiritual quests but are envisioned as the harbingers of a new world order. They possess the ability to bring about a transformation in the collective consciousness of humanity, steering it towards a divine life on Earth. In essence, they serve as the catalytic agents for the collective evolution of the human race, helping to manifest the Divine on the earthly plane.

Sri Aurobindo does not see this transformation as an instantaneous event but as a gradual process where the Supramental principle infuses and uplifts every aspect of human existence, from our physical bodies to our social systems. This transformation will lead to a life on Earth that is characterized by peace, harmony, and a sense of divinity in everyday actions.

In the second part of the book, Sri Aurobindo addresses the question of the why. Why is humankind overcome with ignorance, error and suffering? According to him, the reason lies with the individualization of the soul, its propensity to consider everything in the universe according to its own point of view to the exclusion of all others.

What is the justification for the 'fall'? The plunge into the fragmentation of ordinary life is not 'a blunder and a fall, but a purposeful descent, not a curse, but a divine opportunity'. Brahman is embodied in the lower hemisphere to take part in what Sri Aurobindo calls 'the adventure of consciousness and joy'.

The last part of the book is devoted to the conditions enabling the advent of the divine life. Brahman's plunge into fragmented matter must be followed by its re-emergence into the Light, its full recovery of Knowledge.

But this is not intellectual knowledge that can be entirely assimilated in our present mould of consciousness—this is supra-intellectual spiritual knowledge that can only come through profound inner experience and a self-becoming, and can best be assimilated and lived in a consciousness widened and deepened by intricate and arduous processes of what is generally known as Yoga.

Yoga, in this sense, may be understood as an accelerated or intensified evolution of consciousness, an evolution which is no longer nature-driven but collaborative—an evolution in which humanity can consciously and wilfully participate with nature in her endless quest for perfection. In other words, a 'conscious evolution'. ■■■

The future of humanity in an AI-dominated world is complex and multifaceted. The concept of spiritual conscious evolution provides a framework for approaching this future with intentionality, ethical consideration, and a sense of collective, inclusive responsibility.

APPENDIX – 1

Constituent Assembly of India Volume IX

__________ Entry 73-A

Mr. President: Then we come to entry 73A, and the amendment in the name of Dr. Diwakar. I take it that it is

not moved. Then there is entry 73-A of Mr. Kamath – Inter-planetary travel.

Shri T.T. Krishnamachari: Sir, may I point out that if we talk about a provision for inter-planetary travel, it would be reducing the proceedings of the House to absurdity.

Shri H.V. Kamath: Sir, I am sure if my Friend Mr. Krishnamachari had kept himself abreast of the advance of science and not busied himself only with trade and commerce, he would not have made such a remark.

Mr. President: Have we reached a stage when the control of inter-planetary travel is necessary?

Shri H.V. Kamath: Yes, Sir, as I will try to show, in a few minutes. Sir, I move that................

Shri T.T. Krishnamachari: Sir, I rise to a point of order. It is an impracticable proposition that my friend is suggesting, and therefore it should not be moved.

Shri H.V. Kamath: After you have heard me, Sir, you may decide.

Mr. President: I will hear him first.

Shri H.V. Kamath: Sir, I beg to move:

"That with reference to amendment No.42 of List I (Sixth Week), after entry 73 of List I, the following new entry be added:..."

(Laughter) Sir, Members of the House are welcome to laugh. But fifty years ago, if anybody, had talked of radios and wireless sets, he would have been held up to derision and mocked at, and perhaps stoned. But today radios and wireless sets have become a matter of course, and of everyday occurrence. I am sure Mr. Krishnamachari has got a wireless set of his own in his house. And it is even supposed to a mark of culture today to have a radio set in every house. Take television. Twenty years ago, perhaps television would have been looked upon as an impossibility. But today in America television has become so very common that important meetings and lectures are televised and shown all over the country. With the rapid advance of science for which this century is famous – I am sure within the last fifty years there has been more advance in the various fields of science than in the previous five hundred years – what with research in X-rays, medicine, jet propelled planes of which we hear so much today, we can expect many big changes in the near future. The advance has been remarkable, phenomenal, if I may use such a word. It was only the other day I read in an American paper – it was I think the *New York Times* – there was a news item that a Company had been established or floated in the United States where applications for journeys to the moon had been invited. It was in dead earnest,– I am not referring to it as a jest. They hope to do the journey probably by rockets. Till a few years ago –

Shri R.K. Sidhva (C.P. & Berar): Can you show me the paper?

Shri H.V. Kamath: Yes, if you will kindly come to my place.

Mr. President: I thought people go to the Chandralok after death.

Shri H.V. Kamath: Yes, Sir, I was having it in mind. The *Gita* itself does say.......

Tatra Chandramasam Jyotiryogi Prapya Nivartate

I do not dispute the possibility of a Yogi going even bodily to the moon by the power of his Siddhis and coming back too. But apart from that, Sir, this has come within the range of possibility, and in a few years' time, it is quite possible that there may be journeys to the moon, and the phrase "man on the moon" will lose all its significance.....

The Honourable Dr. B.R. Ambedkar: I don't quite understand whether the proposals of my friend relate to matters which are unknowable or which relate to matters which are unknown. If they are unknown, then we have wasted our time. But if they are unknown and not unknowable, then we have enough powers to deal with them. Why bother with any entry at all?

Mr. President: I will put Mr. Naziruddin Ahmad's amendment to the vote. The question is:

"That in List III (Sixth Week), with reference to amendment No.227 in the proposed new entry 73-A the following be added at the end: 'travel between the planets and the satellites and between the satellites.'"

APPENDIX – 2

An Alternate Vision for Society—Auroville

In the heart of Tamil Nadu, India, lies a city unlike any other—Auroville. Dreamed into existence by Sri Aurobindo and his spiritual collaborator, Mirra Alfassa, also known as the 'Mother', Auroville serves as a living testament to their vision for a global village. It aims to be a place where people from around the world can come together, transcending nationality, religion, and other social constructs, to seek divine consciousness and the unity of humankind.

Sri Aurobindo was a twentieth-century philosopher, yogi, and nationalist who had a tremendous influence on India's independence movement and spiritual landscape. His works delve deeply into the nature of human consciousness and its evolution, providing an integrative worldview that bridges the East and the West.

The Mother, born in Paris and closely associated with Sri Aurobindo, was a spiritual luminary in her own right. She moved to India and collaborated with Sri Aurobindo, contributing significantly to the Sri Aurobindo Ashram in Pondicherry and later establishing Auroville.

At the heart of Auroville lies the Matrimandir, a golden, spherical building that serves as a space for individual silent concentration. The city itself is designed to be a self-sustaining community. It aims to combine the best aspects of various forms of governance, economic systems, and social structures to form a harmonious society.

The community is built on the principles of unity, sustainability, and integral living. Inspired by Sri Aurobindo's Integral Yoga, the lifestyle in Auroville is geared toward achieving a higher state of consciousness while living in harmony with the Earth and each other.

Some argue that the vision of Auroville is a utopian dream, unrealistic and unattainable. Yet, the existence of this city and its continued growth over the years proves otherwise. Auroville has indeed faced challenges, ranging from internal governance issues to external scepticism, but it has endured, attracting thousands of residents from across the globe and becoming a unique experiment in human unity and sustainable living.

Sri Aurobindo and the Mother's vision for Auroville serves as a beacon, inviting us to imagine a world where divisions dissolve and humanity lives as one family. It encourages us to strive for a higher, more inclusive form of consciousness, one that embraces diversity while recognising the underlying unity of all. As we navigate through a world marred by divisiveness, the model of Auroville offers an alternative, a glimpse of what could be possible if we dare to reach for unity in diversity, guided by a higher consciousness.

Thus, Auroville stands not just as a physical location but as an enduring symbol of what humanity can achieve when guided by a vision that transcends our limitations and seeks to fulfil the deepest yearnings of the human soul.

APPENDIX – 3

Bird's Eye Perspectives on some Transformational Technologies (Apart from AI)

The Internet of Things (IoT)

The IoT refers to the network of interconnected devices embedded with sensors, software, and connectivity, allowing them to collect and exchange data. IoT has the potential to revolutionise sectors like healthcare, agriculture, smart homes, and industrial automation.

Blockchain and Distributed Ledger Technology (DLT)

Blockchain offers secure and transparent decentralised systems for recording transactions and managing digital assets. It has the potential to revolutionise finance, supply chain management, healthcare, and other industries by enhancing security, efficiency, and trust.

Robotics and Automation

Robotics and automation technologies are transforming industries like manufacturing, logistics, and healthcare. Collaborative robots, known as cobots, are working alongside humans, enhancing productivity, safety, and efficiency.

5G and Next-Generation Connectivity

The widespread adoption of 5G networks will bring ultra-fast Internet speeds, low latency, and high device connectivity. It will enable advancements in autonomous vehicles, remote surgery, augmented reality (AR), virtual reality (VR), and smart cities.

Renewable Energy and Storage

Battery power shall drive the world (pun intended). Already ubiquitous in all our devices—from cell phones to laptops—and now with cars, scooters, buses, and railways going fully electric at an ever-accelerating pace, the humble battery and fuel cell shall rule the roost.

As the world focuses on combating climate change, renewable energy sources like solar and wind power are becoming more prevalent. Advancements in energy storage technologies, such as batteries and fuel cells, are crucial for storing and utilising renewable energy effectively.

Biotechnology and Genetic Engineering

Advances in biotechnology, such as gene editing tools like CRISPR, are revolutionising healthcare, agriculture, and environmental sustainability. It holds the potential to cure genetic diseases, develop personalised medicine, and enhance crop yields.

Edge Computing

Edge computing involves processing and analysing data closer to the source or at the 'edge' of the network rather than relying solely on cloud-based services. It enables faster response times, reduces network congestion, and enhances data privacy. Edge computing is particularly relevant for applications like autonomous vehicles, IoT, and real-time analytics.

Quantum Networking & Cryptography

Quantum cryptography utilises the principles of quantum mechanics to secure communication channels by detecting any tampering or eavesdropping attempts. It offers enhanced data security and privacy, protecting sensitive information in fields like finance, government, and healthcare.

Quantum networking aims to develop secure communication networks based on quantum principles, enabling unhackable and highly secure data transmission. It has potential applications in cryptography, financial systems, and secure communication.

Synthetic Biology

Synthetic biology combines biology, genetics, and engineering to design and construct new biological parts, devices, and systems. It has the potential to create biofuels, new materials, sustainable agriculture, and personalised medicine.

Drones and Autonomous Vehicles

Autonomous vehicles, including self-driving cars, drones, and unmanned aerial vehicles (UAVs), are expected to transform transportation and logistics. They have the potential to improve road safety, reduce traffic congestion, and enhance efficiency.

Brain-Computer Interfaces (BCIs)

BCIs establish a direct communication pathway between the brain and external devices, allowing for the control of machines or the exchange of information through thoughts. BCIs have the potential to assist people with disabilities, enhance human cognition, and revolutionise human-computer interaction.

Space Exploration and Colonisation

With the renewed interest in space exploration, including government and private initiatives, there is a growing focus on technologies related to space travel, satellite deployment, asteroid mining, and establishing sustainable human colonies on other planets.

3D Printing and Additive Manufacturing

3D printing enables the creation of three-dimensional objects by depositing materials layer by layer. It has the potential to revolutionise manufacturing, allowing for customisation, rapid prototyping, and decentralised production.

Nanotechnology

Nanotechnology involves manipulating matter at the nanoscale level, enabling the creation of materials and devices with unique properties. It has applications in medicine, electronics, energy, and environmental sustainability.

Smart Materials

Smart materials can change their properties in response to external stimuli, such as temperature, light, or pressure. They have applications in various fields, including aerospace, construction, healthcare, and electronics.

Internet of Medical Things (IoMT)

IoMT refers to the network of interconnected medical devices, wearables, and sensors that collect and exchange health-related data. It has the potential to revolutionise healthcare delivery, remote patient monitoring, and personalised medicine.

Precision Agriculture

Precision agriculture leverages technologies like drones, sensors, and data analytics to optimise agricultural practices. It enables farmers to make data-driven decisions, improve crop yields, conserve resources, and minimise environmental impact.

Surveillance and Big Data

1984 is finally here, and Big Brother is always watching.

Biometric technologies use unique physical or behavioural characteristics, such as fingerprints, iris patterns, or facial features, for identification and authentication purposes. They have applications in security systems, access control, and mobile devices.

About the Author

Kunal Nandwani is a serial tech entrepreneur and co-founder and CEO of uTrade and Cove Identity. He is also an active angel investor.

Prior to that, Kunal worked with Lehman Brothers, Nomura and BNP Paribas in London. He holds a bachelor's degree in computer science from PEC (India) and an MBA from ESSEC (Paris).

He has authored two prior books—*Squaring the Blockchain Circle* in 2018, followed by *Sociopreneur: Zero to One* in 2020. *Techtopia* is his third book.